Library Ref: P04COMP0046
Library No: 2644.

Issued by: DOSG Library
Abbey Wood

Date: 04/02/05.

Holder: HOING.

Evaluating the Measurement Uncertainty
Fundamentals and Practical Guidance

Series in Measurement and Technology

Series Editor: **M Afsar**, Tufts University, USA

The Series in Measurement Science and Technology includes books that emphasize the fundamental importance of measurement in all branches of science and engineering. The series contains texts on all aspects of the theory and practice of measurements, including related instruments, techniques, hardware, software, data acquisition and analysis systems. The books are aimed at graduate students and researchers in physics, engineering and the applied sciences.

Other books in the series

Uncertainty, Calibration and Probability:
The Statistics of Scientific and Industrial Measurement
C F Dietrich

SQUIDs, the Josephson Effects and Superconducting Electronics
J C Gallop

Physical Acoustics and Metrology of Fluids
J P M Trusler

Forthcoming titles

Density Measurement and Hydrometry
S V Gupta

Series in Measurement Science and Technology

Evaluating the Measurement Uncertainty

Fundamentals and Practical Guidance

Ignacio Lira

Pontificia Universidad Católica, Chile

With an introductory chapter by T J Quinn, Director of the BIPM

Institute of Physics Publishing
Bristol and Philadelphia

© IOP Publishing Ltd 2002

All rights reserved. No part of this publication may be reproduced, stored in a retrieval system or transmitted in any form or by any means, electronic, mechanical, photocopying, recording or otherwise, without the prior permission of the publisher. Multiple copying is permitted in accordance with the terms of licences issued by the Copyright Licensing Agency under the terms of its agreement with Universities UK (UUK).

British Library Cataloguing-in-Publication Data

A catalogue record for this book is available from the British Library.

ISBN 0 7503 0840 0

Library of Congress Cataloging-in-Publication Data are available

Series Editor: **M Afsar**, Tufts University, USA

Commissioning Editor: John Navas
Production Editor: Simon Laurenson
Production Control: Sarah Plenty
Cover Design: Victoria Le Billon
Marketing Executive: Laura Serratrice

Published by Institute of Physics Publishing, wholly owned by The Institute of Physics, London

Institute of Physics Publishing, Dirac House, Temple Back, Bristol BS1 6BE, UK

US Office: Institute of Physics Publishing, The Public Ledger Building, Suite 1035, 150 South Independence Mall West, Philadelphia, PA 19106, USA

Typeset in the UK by Text 2 Text, Torquay, Devon
Printed in the UK by MPG Books Ltd, Bodmin, Cornwall

*To my daughters,
Andrea and Margy*

Uncertainty in the presence of vivid hopes and fears is painful, but must be endured if we wish to live without the support of comforting fairy tales.
Bertrand Russell

Contents

	Preface	xi
1	**Metrology, its role in today's world**	**1**
1.1	The Convention du Mètre and SI units	3
1.2	Metrology in high-technology manufacturing industries	5
	1.2.1 Precision of manufacture and performance	5
	1.2.2 Improvements in industrial measurement accuracy	7
	1.2.3 Direct access to SI units in manufacturing industry	8
	1.2.4 Application of accurate time measurement	9
	1.2.5 Measurement in the Joint European Torus equipment	11
	1.2.6 Summary	11
1.3	The need for accuracy rather than simply reproducibility	11
1.4	Metrology in human health and in environmental protection	13
1.5	Metrology and fundamental physics	15
	1.5.1 Testing of physical theory	16
	1.5.2 The fundamental physical constants	17
1.6	The role of national and regional metrology organizations	18
	1.6.1 The origins of national metrology laboratories	18
	1.6.2 The need for a research base in measurement science	19
	1.6.3 International representation	20
	1.6.4 International cooperation between metrology laboratories	20
1.7	The cost of maintaining a national measurement system	22
1.8	Final remarks—a personal view	22
	References to chapter 1	23
2	**Basic concepts and definitions**	**25**
2.1	An example: tolerance checking	26
2.2	Values and units of measurement	28
2.3	Calibration and traceability	30
2.4	Measurement uncertainty	32
2.5	Measurement error, accuracy and precision	35
2.6	The measurement model	39
2.7	Some additional remarks	43

3 Evaluating the standard uncertainty — 45
- 3.1 Quantities as random variables — 47
- 3.2 Standard, mutual and expanded uncertainties — 49
  - 3.2.1 Best estimate and standard uncertainty — 50
  - 3.2.2 Mutual uncertainties — 52
  - 3.2.3 The expanded uncertainty — 53
- 3.3 Some useful pdfs — 54
  - 3.3.1 The normal pdf — 54
  - 3.3.2 The trapezoidal, rectangular and triangular pdfs — 56
  - 3.3.3 The logarithmic pdf — 58
- 3.4 The law of propagation of uncertainties (LPU) — 61
- 3.5 The special case of only one input quantity — 66
  - 3.5.1 Second-order Taylor series expansion — 67
  - 3.5.2 Obtaining the expectation and variance directly — 69
  - 3.5.3 Obtaining the pdf: Dirac's delta function — 70
- 3.6 Monte Carlo methods — 75
- 3.7 Imported primary quantities — 76
- 3.8 Directly measured primary quantities: type A analysis — 81
- 3.9 Digital display devices — 85
- 3.10 Correction quantities: systematic effects — 88
- 3.11 Correction quantities: calibration — 92
  - 3.11.1 Calibration of material measures — 93
  - 3.11.2 Calibration of measuring instruments — 96
  - 3.11.3 Calibration curve — 96
  - 3.11.4 Verification and adjustment — 98

4 Evaluating the expanded uncertainty — 102
- 4.1 Coverage factor and coverage probability — 103
  - 4.1.1 The normal pdf — 103
  - 4.1.2 Trapezoidal, rectangular and triangular pdfs — 104
  - 4.1.3 Student's-t pdf — 105
- 4.2 Coverage intervals and coverage uncertainties — 107
- 4.3 The consumer's and producer's risks — 109
- 4.4 The acceptance interval: guardbanding — 110
- 4.5 Type A analysis revisited — 111
- 4.6 Quantities evaluated from a measurement model — 114
- 4.7 Standard *versus* expanded uncertainty — 117

5 The joint treatment of several measurands — 119
- 5.1 Vectors and matrices — 120
- 5.2 The generalized law of propagation of uncertainties (GLPU) — 123
  - 5.2.1 The Newton–Raphson method — 124
  - 5.2.2 Linear models — 125
- 5.3 The input uncertainty matrix — 127
- 5.4 Applications of the generalized law of propagation — 127

			Contents	ix

	5.5	Least-squares adjustment	133
		5.5.1 Classical least-squares	134
		5.5.2 Least-squares under uncertainty	135
		5.5.3 Linear models	137
	5.6	Curve-fitting	138
		5.6.1 Solution using the generalized formalism	139
		5.6.2 An alternative solution	142
	5.7	Consistency analysis	144
	5.8	Applications of least-squares adjustment	149
		5.8.1 Reproducible measurements	149
		5.8.2 Calibration of a thermocouple	150
		5.8.3 Difference comparison measurements revisited	155
		5.8.4 Calibration of a balance	160
6	**Bayesian inference**		**166**
	6.1	Bayes' theorem (BT)	167
	6.2	BT in terms of probability density functions	168
	6.3	Example	170
	6.4	Non-informative priors	173
	6.5	Repeated measurements	175
		6.5.1 Normal probability model	176
		6.5.2 Rectangular probability model	180
		6.5.3 Digital indications	182
		6.5.4 Reproducible measurements	184
	6.6	The principle of maximum entropy (PME)	187
		6.6.1 Range and estimate given	189
		6.6.2 Range, estimate and uncertainty given	190
		6.6.3 Estimate and uncertainty given	192
	6.7	Evaluation of the producer's and consumer's risks	192
		6.7.1 The Bayesian viewpoint	193
		6.7.2 The frequency-based viewpoint	196
		6.7.3 The risks in instrument verification	198
	6.8	The model prior	201
	6.9	Bayesian analysis of the model $Z = \mathcal{M}(X, Y)$	202
		6.9.1 The model $Z = X + Y$	203
		6.9.2 The models $Z = X/Y$ and $Z = XY$	207
		6.9.3 The model $Z = X + Y$ with prior knowledge about Z	208
		6.9.4 Known variance of the observations of Y	219
		6.9.5 Summary and conclusions	220
	Appendix: Probability tables		**222**
	Glossary		**230**
	References		**236**
	Index to chapters 2–6		**239**

Preface

Measurements are ubiquitous. Because they are indispensable for maintaining the diverse functions of modern technology-based society, they are constantly carried out everywhere. Their applications range from the complicated experiments needed to develop and test the newest scientific theories to everyday uses in industry, in commerce and in our homes. And yet, unbeknownst to most people, the information provided by measurement can rarely be assumed to be complete. Indeed, any measurement result is in general just a *point estimate* of the measured quantity, the 'true' value of which remains unknown. For this reason, one should appraise the dispersion of the values about the estimate that could as well be attributed to the measurand. In metrology—the science of measurement—this parameter is called the *measurement uncertainty*. If it is stated the user can assess the confidence to be placed on the measurement result. Unfortunately, there is no unique way to express quantitatively the 'doubt' that the uncertainty represents. As a consequence, different and in some cases conflicting uncertainty evaluation procedures were developed over the years.

In recognition of this lack of accord, in 1978 the *Comité International des Poids et Mesures* (CIPM) requested its executive body, the *Bureau International des Poids et Mesures* (BIPM), to formulate agreed-upon fundamental principles on the problem of uncertainty evaluation. Together with experts from 11 National Metrology Institutes, the BIPM produced Recommendation INC-1 (1980). The task of elaborating the principles declared therein into a detailed practical guide was assigned to the Technical Advisory Group on Metrology (TAG 4) of the International Organization for Standardization (ISO), which in turn convened a Working Group (WG 3) composed of experts nominated by the BIPM, the International Electrotechnical Commission (IEC), the International Organization for Standardization (ISO) and the International Organization of Legal Metrology (OIML). The outcome was the *Guide to the Expression of Uncertainty in Measurement*, published by ISO in the name of the BIPM, the IEC, the OIML, the International Federation of Clinical Chemistry (IFCC), the International Union of Pure and Applied Chemistry (IUPAC) and the International Union of Pure and Applied Physics (IUPAP). That authoritative document is known popularly (but not very elegantly) as the GUM. In this book it will be referred to as the *Guide*.

The *Guide* was intended to orient metrologists into the complexities of the evaluation of uncertainty. In that sense, its anonymous authors were very successful. It won rapid acceptance, and today it is considered by most if not all professional metrologists as their 'uncertainty Bible'. Its publication has promoted the achievement of the international consensus that now exists on the practice of stating formally measurement results, and has made possible the international comparison of those results, especially in regard to standardization, calibration, laboratory accreditation and metrology services. Moreover, metrological requirements stated in quality-related documentary standards, such as ISO 9000 and ISO 17025, implicitly call for all measurement evaluation tasks to be performed in accordance with its recommendations. Likewise, other standards and guides, such as EA-4/02 *Expression of the Uncertainty of Measurement in Calibration* (by the European Cooperation for Accreditation) and NIST TN 1297 *Guidelines for Evaluating and Expressing the Uncertainty of NIST Measurement Results* (by B Taylor and C Kuyatt) are strictly based on the *Guide*.

The problem of performing an uncertainty budget is actually quite simple in most cases. Its solution amounts only to formulating a proper measurement model, and to evaluating that model in accordance with the rules given in the *Guide*. Nevertheless, as technical manager of a calibration laboratory in dimensional metrology, I am all too aware of the fact that, when confronted with the task, most practitioners still feel that they enter uncharted territory. This happens probably because, as its title implies, the *Guide* is not a textbook. It is a rather terse and not very user-friendly document. Its main body is short and does not dwell much on theoretical aspects, focusing instead on providing a set of rules that effectively cover most measurement situations likely to be encountered in practice. This approach is fine as long as no additional guidance is required. Otherwise, one has to turn to one of its appendices and hope to find the information there, for no subject index is provided. Moreover, the appendices were clearly written by different people. As a result, they show inconsistencies in the various treatments. More importantly, they do not include some specialized topics for the evaluation of experimental data. For example, the metrologist may need to evaluate a nonlinear model, to perform a least-squares adjustment or to use matrix algebra when several quantities are to be measured through a system of models. In these applications the *Guide* is clearly insufficient.

Furthermore, the *Guide* uses mainly conventional frequency-based statistical theory. Many times this cannot be fully applied, because the metrologist is only rarely in possession of large amounts of measurement data. In these cases a Bayesian analysis is more appropriate. This is based on Bayes' theorem, which is a natural way of conducting inference from the measurement experience and allows for a rigorous way to take into account any prior information about the measured quantity that may exist. Bayesian statistics is well established and is presented in various specialized books. However, they have a strong mathematical flavour and tend to concentrate on issues that are not of much interest to metrologists.

I do not mean to be harsh in my criticism of the *Guide*. As an engineering professor, I feel that it deserves recognition not only within the metrological community, but in academia as well. Even though most of my colleagues acknowledge that measurements are vital, they tend to overlook the fact that one must always work with properly calibrated measuring instruments. As a result, few students realize that experimental data should not be accepted at face value. This state of affairs probably occurs in other technical curricula as well as in scientific disciplines such as biology, chemistry and physics. In some laboratory courses students may learn about the related subject of 'error analysis', for which a number of books are available. However, 'error' and 'uncertainty' are quite different concepts. The former is a point estimate of the difference between the estimated and the 'true' value of the measured quantity, while the latter is a range of possible values. Errors can be measured (indeed, they are when a measuring instrument is calibrated) and therefore they are uncertain. On the contrary, the frequently used notion of an 'estimated' or 'probable' uncertainty constitutes a contradiction in terms.

In summary, the existing international documents on the evaluation of the measurement uncertainty are not self-sufficient, do not quite serve as manuals for self-study, are not sold at regular bookstores and cannot be used effectively by instructors to assist students who face the subject for the first time. Moreover, most available textbooks related to the matter are not addressed to people interested mainly in measurement or their approaches may be at odds with the concepts in the *Guide*. Finally, technical articles on measurement uncertainty that appear periodically in specialized journals, such as *Measurement Science and Technology*, *Metrologia* or *Measurement*, may be hard to follow by readers who do not have the necessary background.

This book is, therefore, an attempt to serve the needs of a very wide audience. It should appeal to metrologists in National Metrology Institutes and calibration laboratories who wish to complement their knowledge of the *Guide*. It should also be useful to teachers and students of applied courses on quality, metrology, instrumentation and related subjects at the undergraduate level in universities and technical schools. It should as well be of much interest to graduate students carrying out experimental research in all fields of science and technology. It should benefit scientists and engineers engaged in research and development, especially in those areas where theories are developed and tested on the basis of measurement data. And it should find a niche in the bookshelves of all professionals that perform or make use of measurements, even those that do not consider themselves 'metrologists'.

The book is organized as follows. Chapters 1 and 2 are introductions, the former to metrology in general and the latter to uncertainty and related concepts. Chapters 3 (on standard uncertainty) and 4 (on expanded uncertainty) form this book's core. They develop systematically most of the points addressed in the *Guide*. By contrast, chapters 5 and 6 are much more specialized; they contain information that cannot be found easily other than in journal articles. Thus, chapter 5 pertains to the case of several quantities measured simultaneously while chapter 6 is devoted to Bayesian statistics.

More specifically, chapter 1—kindly contributed by Dr T J Quinn, Director of the BIPM—sets out to explain why accurate measurements are required, how they are obtained and why the search for ever-increasing accuracy will certainly continue. It presents measurement as an important element in the fabric of industrial society and reviews, in technical and economic terms, the roles played by National Metrology Institutes and measurement systems in high-technology industries, in health and safety, in the protection of the environment and in basic science.

Chapter 2 introduces the concept of measurement uncertainty. It uses as an example a measurement situation typically encountered in industrial settings, where a product characteristic has to be compared with the corresponding specification to determine whether the product should be accepted or rejected. As the example progresses several definitions are given; these are transcribed from the ISO *International Vocabulary of Basic and General Terms in Metrology*, known generally as the VIM. This chapter also stresses the importance of formulating a measurement model as an essential first step, without which the evaluation of uncertainty is a hopeless task.

Chapter 3 is dedicated to the evaluation of the uncertainty associated with the estimate of a quantity that is modelled explicitly in terms of a number of input quantities. It starts by defining the 'standard' uncertainty as the square root of the variance of the associated probability density function (pdf). The concepts of correlation and of expanded uncertainty are also introduced briefly at this point. After studying some 'typical' pdfs, the 'law of propagation of uncertainty' is derived. It applies to models that are linear or not too strongly nonlinear. Next, an analytic procedure applicable to nonlinear models involving a single input quantity is described. Also, a succinct explanation of the Monte Carlo simulation technique is given. The chapter ends by considering the evaluation of the uncertainty associated with the estimates of input quantities that are imported from other external measurement operations, and with those that are measured directly with a measuring instrument. The latter involves, in general, three uncertainty components. The first is associated with the repeatability of the indications and is evaluated with the so-called 'type A' procedure. The other two components are those associated with all applicable systematic effects (including calibration corrections) and with the resolution of the instrument.

Chapter 4 concentrates on the problem of determining an 'expanded' uncertainty. This is defined as a range of values within which one can be almost sure that the true value of the quantity lies. The concept is introduced by first defining the producer's and consumer's risks as the probability of rejecting an acceptable item, and of accepting a non-conforming item, respectively. It is shown next that the consumer's risk is usually reduced in practice by subtracting twice the expanded uncertainty from the tolerance for a given product. The chapter then proceeds to explain the conventional procedure to determine the expanded uncertainty associated with the estimate of a directly measured quantity; and to show how this procedure can be extended to quantities modelled in terms of other quantities. This requires an explanation of the central limit theorem, of the use of the Welch–Satterthwaite formula in conjunction with Student's-t probability density function, and of the difference between confidence level and coverage probability. It is also shown that these conventional tools present some theoretical inconsistencies that render their application questionable.

Chapter 5 deals with the cases of measured quantities that are modelled implicitly, and of several output quantities that are inter-related through a number of models involving common input quantities. It begins with a brief review of the matrix formalism used throughout, thus avoiding the more cumbersome and restricted summation-type formulae. The concepts of uncertainty and sensitivity matrices are then presented, they allow the generalization of the law of propagation of uncertainties. This generalization is illustrated through various examples. Next, the subject of least-squares adjustment is presented, first from a very general point of view and then as it applies to the frequently encountered case of curve adjustment. The important topic of consistency analysis of the input data is also explored. Several worked-out examples are then given to illustrate the least-squares procedure.

Finally, chapter 6 presents the essential tools of Bayesian statistics within the framework of measurement data evaluation. These tools are applied to infer the estimated values and uncertainties of directly measured quantities. It is shown that the procedures (but not necessarily their results) differ radically from those of frequency-based statistical analysis. The principle of maximum entropy is next introduced as a tool to construct pdfs that just agree with the available information. A rigorous procedure to evaluate the consumer's and producer's risks follows. Finally, the concept of model prior is presented and is applied to the evaluation of the expanded uncertainty in some simple measurement models.

Those to whom this book is addressed should not feel discouraged by the mathematics. Although some formulae and derivations will appear to be lengthy or complicated, only a working knowledge of elementary calculus and basic algebra is needed to understand them. Probability and statistical concepts are, of course, used throughout, but they are introduced and explained as their need arises. To facilitate comprehension, I have endeavoured to keep a clear and consistent notation and, where appropriate, to include several figures and numerical examples. Some of these are rather trivial, others are more complex and have been solved with the help of mathematical software packages.

After studying the contents of this book, conscientious readers should attain a reasonable mastery of the subject. Not only are all topics addressed in the *Guide* covered here in a unified manner, but also much more material of interest has been included. Still, in the *Guide* one reads that

> a framework for assessing uncertainty ... cannot substitute for critical thinking, intellectual honesty, and professional skill. The evaluation

of uncertainty is neither a routine task nor a purely mathematical one; it depends on detailed knowledge of the nature of the measurand and of its measurement. The quality and utility of the uncertainty quoted for the result of the measurement therefore ultimately depend on the understanding, critical analysis, and integrity of those who contribute to the assignment of its value.

To this declaration I subscribe fully.

I am indebted to Dr Clemens Elster, from Germany's Physikalisch-Technische Bundesanstalt (PTB), for a careful reading of the manuscript and for making valuable suggestions for its improvement.

Ignacio Lira
December 2001

Chapter 1

Metrology, its role in today's world

Text of a lecture by Dr T J Quinn, Director of the BIPM, given in Tokyo on 14 October 1993 to the Japanese Instrument Manufacturers Federation. Reproduced from Rapport BIPM-94/5 with the permission of Dr Quinn.

In today's society there exists a vast, often invisible, infrastructure of services, supplies, transport and communication networks. Their existence is usually taken for granted but their presence and smooth operation are essential for everyday life. Part of this hidden infrastructure is metrology, the science of measurement.

Confidence in measurement enters into our lives in a multitude of ways. It allows high-volume goods, such as crude oil or natural gas, to be traded in the secure knowledge that the millions of tons of oil or cubic metres of gas bought and sold are correctly measured from the super-tanker to the petrol pump and from the high-pressure cross-border pipeline to the domestic gas meter. It is a truism that accurate measurements are required for the efficient manufacture of components for such varied things as internal combustion and gas turbine engines, where reliability and long life depend upon manufacturing tolerances of micrometres; and in compact disk players, which incorporate lenses to focus laser light that are made to tolerances below a tenth of a micrometre. In terms of high-technology industrial production, the list of applications requiring accurate measurement is endless.

International telecommunication systems work reliably and efficiently, but for high rates of data transmission time scales across the world must be closely coordinated and must not fluctuate from microsecond to microsecond, from minute to minute or from day to day. Within the industrialized world, national time scales are linked to within about a microsecond to UTC, the international time scale based on atomic clocks. UTC is also the time scale maintained by the atomic clocks (corrected for the effects of general relativity) on board the satellites of the Global Positioning System (GPS), the military and civil worldwide navigation system set up by the US Department of Defense.

The practice of medicine requires careful and sometimes difficult measurements both in diagnosis, as in the measurement of blood cholesterol level, and in therapy, as in the measurement of x-ray or γ-ray dose for the treatment of some forms of cancer. In these measurements, reliability is of the utmost importance for errors can, and have been known to, kill.

In agriculture, the testing of food products and the protection of the environment, measurement is becoming increasingly important in providing the basis for, and the means of verifying conformity to, a wide range of legislation. Much of this is related to ensuring that pesticide and heavy-metal residues are kept at safe levels.

Individually each of these applications of measurement and the need to have confidence in them is obvious when they are brought to our attention. It is perhaps less obvious that reliable and accurate measurements having long-term stability can only be ensured by a measurement system firmly linked to fundamental physics.

The continuous search for greater accuracy in measurement is driven by the increasing demands of high technology and its success is made possible by advances in physics which themselves often lead to new technologies. An example of this is the laser. Almost at a stroke, the invention of the laser allowed measurements of optical wavelengths to be made with accuracies some two orders of magnitude higher than was possible before. It also led to the development of instruments and devices, such as the compact disk player, that both incorporate a laser and rely for their manufacture upon the advances in dimensional metrology made possible by the laser.

The fundamental physics upon which all of today's advanced technology is based is itself reliable only to the extent that its predictions can be tested in a quantitative way. The testing of physical theories by the carrying out of increasingly accurate experiments and the establishment of a consistent set of the fundamental physical constants are an essential component of the progress of science and are parts of the activity known as metrology. The provision of reliable and accurate data on the properties of materials is a related activity, and makes an essential contribution to the infrastructure of engineering and high-technology manufacturing.

For the best part of one hundred years, a close network of national metrology laboratories has developed and maintained what is now a worldwide measurement system assuring accuracy and uniformity in measurement standards at both the national and international level. These laboratories carry out long-term basic research on measurement with the aim of advancing the frontiers of measurement science and thereby maintaining their expertise necessary to carry out the function of a national measurement laboratory. Only active involvement in research can give them the competence to carry out high accuracy calibrations of measurement standards, to provide authoritative advice to industry and government and to represent the national interest at the international meetings where binding decisions are taken on measurement and measurement-related questions.

The way in which metrology is organized and the funds assigned to it vary from country to country. Because this activity tends to be both invisible and costly, governments sometimes find difficulty in justifying its support. In those cases where well-established laboratories and national measurement systems exist, operating within a clear framework of national legislation, and where neither the short nor the long-term goals are in question, the problems that arise are those of making the most efficient use of resources. In other cases, where there is no clear understanding in government, or occasionally even in industry, of the role that metrology plays in the industrial and commercial success of the country, the situation is very different. The national laboratory, or whatever organization for metrology exists, is subject to continual review: the need for the work, its basic goals and the means to attain them, as well as every detail of the programme, are permanently in question. Under these circumstances the programme of work is drawn up taking careful account of the costs but with little reference to its value. Only too often the result is that not the government, which pays for the work, nor the industry and commerce that should benefit from it, nor the national laboratory which carries it out, are satisfied or well served.

The aim of this chapter is to outline the basic justification for government support of metrology by demonstrating its role in national and international affairs and, thereby, to provide some answers to those who ask: What is metrology for? Why does it cost so much? Why do we keep having to do it?

1.1 The Convention du Mètre and SI units

The origins of metrology go back to the earliest traders whose instruments of exchange were gold coins and the beam balance. In some respects little has changed since, except that these days the coins are rarely of gold. Then, as now, the buyer and seller of goods had to agree upon the units of exchange. Today, however, we do not worry about fluctuations in the values of, or rates of exchange between, the units of measurement as we still do for values and rates of currency, because we now have an agreed and stable system of units of measurements which is recognized worldwide. This security is based on the Convention du Mètre, signed in Paris in 1875 and maintained ever since by the Bureau International des Poids et Mesures at Sèvres and the national metrology laboratories. The need for such an international convention was first recognized in the 1860s as being one of the essential requirements for the growth of international trade in manufactured and industrial products. The decimal metric system created at the time of the French Revolution was, by the middle of the 19th century, widely used not only in France but also in most European countries. The metric system was thus chosen as the basis for the international system of units created by the Convention which is an intergovernmental treaty between the now 47 adhering nations[†], comprising substantially all of the industrialized nations of the

[†] Forty-nine nations as of 2001.

world. The Convention established a permanent organizational structure within which governments can act in common accord on all matters relating to units of measurement. This includes a requirement for periodic Conferences Gènèrales des Poids et Mesures (CGPMs) that now take place once every four years in Paris, and a Comité International des Poids et Mesures (CIPM) whose membership is 18 individuals elected by the CGPM. The Convention also created the Bureau International des Poids et Mesures (BIPM) which today, with its laboratories and a staff of almost 70 persons, acts as the world centre for metrology.

Soon after the foundation of the BIPM a number of the major trading and industrial nations of the world established their own national laboratories for standards. Among these were the Physikalische-Technische Reichsanstalt, Berlin (1887), the National Physical Laboratory, Teddington (1900), the Bureau of Standards, Washington (1901) and the National Research Laboratory of Metrology, Tokyo (1903). Others soon followed and there now exist metrology laboratories in virtually all of the industrialized nations of the world. These are responsible for providing the national standards of measurement and in many cases for managing the national measurement system, comprising calibration laboratories, certification and accreditation bodies, and legal metrology.

The metric system, upon which the Convention du Mètre was based, grew to include electrical and photometric units and in 1960 the modern form of the metric system, the Système International d'Unités (SI), was formally adopted by the 10th CGPM. The SI now has seven base units [1]‡ and includes derived and supplementary units together with prefixes for the multiples and submultiples of the SI units and has established a comprehensive specification for units of measurement. The SI is now the unit system almost universally used for science and technology. Even in the USA, the last bastion of the imperial system, the move to SI continues. It is impossible for it to be otherwise. Measurement is an intimate part of high-technology industry, and most of the world's major companies are now multi-nationals having production plants in more than one country. The continuing use of multiple systems of units leads to reductions in production efficiency and is a source of errors, thus running counter to today's drive for competitiveness and quality§.

Before the Convention du Mètre the local, regional and national units of measurement, and their enforcement in law, were maintained by what were generally called Weights and Measures offices. The successors of these offices, usually still under the same name, are now responsible for legal metrology. This is the aspect of metrology directly controlled by law and in most countries concerns those measurements that enter into the buying and selling of goods to the public. Very strict requirements govern such transactions since large quantities of goods and correspondingly large amounts of money are involved. For example, each year in the UK, some 30 billion litres of petrol are sold with a value of some

‡ References to this chapter appear on page 23.
§ The failure in 1999 of NASA's Mars Climate Orbiter Mission was due to a confusion in the units of measurement.

15 billion pounds. Overall, about one-third of the gross domestic product of Western industrialized countries is in the form of internal retail trade. All of this comes under the purview of the Weights and Measures offices or their equivalent. These, in turn, obtain their reference standards for the verification of retail measuring instruments from the national metrology laboratories. The accuracies with which retail measurements must be made is not generally high compared with the capabilities of the national laboratories. Their reliability, however, must be beyond reproach and the possibilities for fraud minimized. Public tolerance for false measures is, quite rightly, very low. The error allowed, for example, in the amount of fuel dispensed by a petrol pump, in the UK is normally about 0.3 %, but 0.3 % of £15 billion is still some £50 million!

Similar examples could be given for international trade in such products as crude oil and natural gas where the same principles apply as for internal national trade. The role of metrology in those cases, however, is to provide a secure and accessible base upon which to establish the international legal and regulatory framework.

1.2 Metrology in high-technology manufacturing industries

A much larger role for metrology, and one which extends its capabilities to the limit, is the provision of the accurate measurements required in high-technology manufacturing and telecommunications. They are reviewed in this section.

1.2.1 Precision of manufacture and performance

About half of all manufactured products is accounted for by individual items such as aircraft, motor vehicles and computers, together with their component parts. The other half mostly comprises goods manufactured in bulk. In the USA, the value of discrete items amounts to about 600 billion dollars (more than 10 % of GDP). Of this, about half is in the automobile industry [2], other important sectors being aerospace and instrumentation. For most of these products their performance and perceived quality, and hence their commercial success, is determined by how well they are made. Engineering tolerances, i.e. the amount by which dimensions are permitted to depart from specification, have fallen by a factor of three every ten years since 1960. Such a reduction in manufacturing tolerances pre-supposes corresponding improvements in precision machining and in metrology.

There are two reasons for this improvement of precision in manufacturing industries over the past 30 years. The first is that in traditional mechanical engineering, improvements in performance and reliability have only been possible through improved precision in manufacture. The second is that many of the new technologies, based upon the practical application of recent discoveries in physics, simply do not work unless high-precision manufacturing is available. Examples of some of these are the electro-optic industries using lasers and fibre optics, the manufacture of large-scale integrated circuits, and the commercial production of positioning and navigation systems using signals from atomic clocks on satellites. Dimensional tolerances of 0.1 μm and below, and timing tolerances of a few nanoseconds, are required in the examples given. Such fine tolerances in manufacture and use require an accuracy in measurement capability at an even finer level.

It is now well known that the performance of many manufactured products continues to improve when the mechanical and other errors of manufacture are reduced far beyond what, at first sight, is required. A striking example of this is to be found in gear boxes of the sort used in gas turbines to drive contra-rotating propellers, an engine configuration now considered to be the most efficient for large passenger aircraft [3]. The capacity of the gear box, defined as the torque that can be transmitted per unit weight, increases dramatically as the individual gear-tooth error falls. Capacity doubled when the tooth position and form error fell from 10 μm to 3 μm, the best at present available in production. A further doubling of capacity is expected when the overall tooth error is reduced from 3 μm to 1 μm. The measurement of position and form of gear teeth with an accuracy of 1 μm is not yet possible either on the production line or in national metrology laboratories. The lifetimes of such components also increase nonlinearly with improved precision of manufacture. The tolerance now set in Japan in the production of pistons for some automobile engines is about 7 μm, similar to that required in the manufacture of mechanical watches!

An example of how quality and commercial success are linked to manufacturing precision may be found in the forces required to open US and Japanese car doors [2]. In the 1980s the forces required to open the doors of Japanese and US cars differed by a factor of three. The origin of the difference was that Japanese cars had a tolerance of 1 mm on door and door assemblies and were therefore easier to open than were American cars where the equivalent tolerance was 2 mm. Ease in the opening of the door is an important factor in the perceived quality of the whole product, and is directly related to precision of manufacture. It had large economic consequences for the US motor industry.

In the manufacture of integrated circuits, the essential component of computers and microprocessors, photolithography is the key technology that allows the production of the millions of individual features present on a typical 22 mm × 15 mm chip. The market for integrated circuits is worth hundreds of billions of dollars worldwide and goes to the manufacturers that can pack the most components on a single chip. It is at present concentrated in a small number of US and Japanese companies. A most demanding dimensional measurement appears in the so-called 'step and repeat' machines that manipulate the photolithographic masks used to put down the successive layers that make up the microcircuits. In these machines, which cost between 2 and 3 million US dollars each, and for which the annual world market is upwards of 500, errors in the optical images of the masks is kept below 60 nm (10^{-9} m) over dimensions of a few centimetres. In today's most advanced machines, images are formed using ultraviolet light from Kr–F excimer lasers at a wavelength of 248 nm. Intense research efforts are being made to allow the use of x-rays or electrons so that even finer resolution, and hence denser packing of components, can be achieved. Associated with the step and repeat machines are high-speed, fully-automated x–y measuring systems for checking the shape and dimensions of photolithographic masks and the finished wafers. These systems have a repeatability over 24 hours of 10 nm in a 200 mm × 200 mm range and incorporate a wavelength stabilized He–Ne laser and active compensation for thermal and mechanical deformation of the silicon wafer.

1.2.2 Improvements in industrial measurement accuracy

To achieve high-precision measurement in production engineering more is required than a simple upgrading of existing precision methods. Among recent initiatives have been the development of new measuring tools, notably the three-coordinate measuring machine (CMM), the application of computers to the calibration and use of such new machines and the provision of direct access to the SI unit of length through laser interferometry.

Three-coordinate measuring machines comprise three mutually perpendicular axes of linear measurement and sophisticated computer software to allow accurate and rapid evaluation of the shape and dimensions of complex pieces such as bevelled-tooth gears, crankshafts or turbine blades. Three-dimensional systematic-error compensation is obtained by calibration using carefully designed test objects. The accuracies of CMMs calibrated in this way can reach a level of about 2 µm in all three coordinate axes over volumes of about 1 m^3. Their performance, however, is critically dependent on the computer software being free of errors. The checking of this software is the most difficult part of the calibration process.

Displacement measurement using laser interferometers can achieve very high accuracy [4] but great care must be exercised to avoid the class of errors known as Abbe errors. These arise essentially from incorrect alignment, so that the optical displacement measured with the interferometer is proportional to, but not equal to, the displacement of the object. Similar errors can arise if an object is displaced by pushing it against a frictional resistance: unless the direction of the applied force is parallel to and coincident with the net resistance vector, the object experiences a torsion stress and deforms. These considerations are important not only in machine tool design, where they have been well known for a hundred years, but also in the design of systems in which the measurements of displacements and dimensions are at the nanometre level and are well known to the designers of the step and repeat machines. Of itself, extreme sensitivity is not a guarantee of accuracy: proper design practice must also be adopted. One of the important activities of the national laboratories in the new field of nanometrology [5], metrology at the scale of nanometres (10^{-9} m), is the exploration of machine behaviour so that good measurement practice can be developed.

1.2.3 Direct access to SI units in manufacturing industry

Direct access to SI units and to quantum-based standards has led to improvements in measurement accuracy in many fields of measurement. The application of optical interferometry, using known wavelengths of light, to dimensional metrology allows accurate measurement standards to be built into three- and two-dimensional CMMs so that intrinsic errors are now limited only by the refractive index of air. In electrical metrology 1 V or 10 V Josephson-array systems allow voltage measurements to be made with reproducibilities of a few parts in 10^{10}. In time measurement, commercial caesium atomic beam clocks allow time to be measured to accuracies of a few parts in 10^{13}, equivalent to a few nanoseconds per day. The possibility of making such measurements either on the production line, as in the case of the voltage measurements, or in practical commercial applications, as in the case of atomic clocks, could not be envisaged if the practical measurements were linked to national standards by the traditional hierarchical chain of calibrations.

Before direct access to these inherently stable units was available in industrial applications, a hierarchical chain of calibrations was the only way to link metrology in manufacture to national and international standards. Consequently, the accuracy of industrial measurement standards was significantly lower than that of national standards. A factor of three or four in each link in the chain could hardly be avoided, so that national standards had to be a factor of ten, and sometimes factors of 20 to 50, more accurate than industrial requirements. Some of the accuracies now required in high-technology manufacture would, if the same scheme applied, call for accuracies in national laboratories well beyond what is presently feasible. For example, the present tolerances of 60 nm in the dimensions of photolithographic masks would, if obtained through a traditional calibration chain, require national dimensional standards of better than 3 nm, a figure which at present is well beyond what is feasible.

In a very real sense, the accurate measurements now needed in certain high-technology industries are at the frontiers of metrology and directly based upon fundamental physics. This new situation is reinforcing the role of national metrology institutes as centres of excellence and expertise in advanced measurement since their advice and services are increasingly called upon by those whose requirements are the most demanding. The practical use of Josephson arrays and atomic clocks for accurate measurement and the development of a full physical understanding of their behaviour was carried out in national metrology laboratories. It was from them that industrial enterprises obtained the basic knowledge that led to commercial applications of these devices. The new field of nanotechnology relies heavily on corresponding research in nanometrology, much of which is being carried out in national metrology institutes [6].

The industrial application of 1 V or 10 V Josephson systems highlights some of the unexpected advantages to be gained by a measurement capability two orders of magnitude in advance of apparent needs. In the commercial production of digital multimeters two parameters are fundamental: linearity and full-scale accuracy. The ability to make measurements on the production line with an accuracy two orders of magnitude better than the final specification of the instruments has at least two important advantages: (i) deviations from mean production specifications are noticed well before they become significant and corrections can be applied (this allows 100 % of the production to be well within specification); and (ii) final calibrations for linearity and accuracy can be made quickly and efficiently and no significant error comes from the calibration equipment. The advantages are thus those of efficiency in production, and quality in the final product. According to the manufacturers now using such systems, these advantages outweigh, by a considerable margin, the extra cost of installing and maintaining the Josephson system.

Although this example is from the field of electrical measurements, it is evident that in most industrial production rapid and accurate measurement systems have similar advantages. Indeed, one of the central concepts of so-called computer-aided flexible manufacturing systems is that accurate measurements made in real time during production allow departures from specification to be corrected before they become significant. Older methods, in which components are manufactured and subsequently tested for conformity to specification with the rejection of a certain fraction of the product, cannot compete with those in which 100 % of the product is within specification. To achieve this, however, measurements taken during manufacture must be made at a level significantly below final tolerance.

1.2.4 Application of accurate time measurement

A different type of requirement for accurate measurement is in systems for global positioning and surveying using time signals from earth satellites. Such systems only become of practical value when the accuracy of the key measurement, in this case time, reaches a certain threshold.

The Global Positioning System (GPS) comprises a constellation of 24 satellites orbiting the earth at an altitude of about 20 000 km. Each has on-board caesium beam atomic clocks and transmits coded time signals that can be picked up by small receivers anywhere on the surface of the earth. The clocks on the satellites are all set to the same time, to within about 30 ns. The principle of GPS positioning calls for the simultaneous observation of four satellites in different parts of the sky and measurement of the apparent time differences between them. The position of the receiver on the ground can be found from these apparent time differences knowing the positions of the satellites at the time the observations were made. The satellites are sufficiently high for atmospheric drag to be negligible and are in sidereal orbits so that their positions can easily be programmed into the receiver memory. The accuracy of the satellite position in space must be at least as good as the required position accuracy, namely about 10 m. In practice, the overall accuracy for real-time position determination is not as good as this and is only about 30 m, due principally to the difficulty of making accurate corrections for ionospheric delays in signal propagation. Furthermore, for civil use, an intentional degradation, known as Selective Availability (SA), is introduced by the US Department of Defense and the accuracy available in the presence of SA is about 100 m. This is of course perfectly adequate for civil navigation. For surveying and other purposes, for which immediate knowledge of the result is not necessary, or for differential measurements, in which a receiver is also placed at a known reference point, the accuracy can be much greater, about 0.1 m. The cost of setting up and maintaining a system such as GPS is very high: each satellite has a life of about eight years and to complete the constellation of 24 satellites has cost more than 10 billion dollars, the second most expensive space programme after the manned landing on the moon. This whole system relies completely on the performance of the clocks.

In addition to the military users of GPS there are many civil users. The commercial applications of worldwide positioning and navigation to an accuracy of 100 m are very wide and there are now many producers of GPS receiving equipment. It is estimated that in 1992 the sales of GPS hardware reached 120 million dollars and the new industry based on GPS and producing digital maps and associated navigation systems had sales approaching 2 billion dollars. Within a few years civil aviation navigation is sure to be based on GPS. New generations of atomic clocks, having accuracies perhaps two orders of magnitude better than those now available, will undoubtedly lead to corresponding improvements in global positioning and corresponding commercial applications. These new clocks, based upon trapped or cooled atoms or ions, are now under development in national laboratories and are expected to have accuracies of parts in 10^{16}.

The technology of accurate time dissemination and the maintenance of national time scales to within a few microseconds worldwide are of high commercial importance [7], quite apart from their use in positioning. Data networks and other telecommunications systems are under continuous pressure to increase the rate of information flow. One limitation to the speed of operation of such systems is jitter in the basic frequency and phase of interconnected parts of the system. If, as is now often the case, different national networks are connected it is a basic requirement that national time and frequency systems fit together without significant jitter. This is the origin of the growing demand to link national time services to within 100 ns, a factor of ten better than what is achieved today.

Quite apart from the technological and commercial applications of GPS for time-scale comparisons and positioning, other unexpected scientific applications are beginning to appear now that the scientific community has become aware of the full potential of the system. For time-scale comparisons and positioning, the delays introduced to the GPS signals by their passage through the ionosphere must be corrected for. Accurate measurement of these delays, however, can provide information on the atmospheric composition and temperature that are of interest to climatologists. Recent studies [8], have indicated that if the ionospheric delays are measured with an accuracy of about 3 ns, equivalent to 1 m of path, which is already possible, GPS could become a valuable tool for the study of long-term changes in atmospheric water and carbon dioxide content as well as temperature in the upper atmosphere.

A substantial increase in accuracy in the measurement of any important physical quantity almost always leads to unexpected applications in fields quite different from that for which the advance was made.

1.2.5 Measurement in the Joint European Torus equipment

A further example illustrating the key role of measurement, and one that can easily be costed, is research aimed at producing power from nuclear fusion. The possible commercial production of energy from controlled nuclear fusion depends on the results of large and expensive experimental devices such as JET (Joint European Torus). This device has so far cost about 1000 Mecus of which 10% has been for the measurement systems. Accurate measurement of the temperature and density of the hot plasma (at about 100 million °C) is crucial to the understanding of the various loss mechanisms that impede fusion. It is evident that with so much at stake the measurement process must be fully understood and a proper evaluation of the uncertainties made. Were this not the case, there is a risk either of unwarranted euphoria, with subsequent disappointment (as has already happened with earlier devices) or of failure to appreciate successes achieved. In this work the cost of poor measurement would be very large and hence the great effort and large amount of money quite rightly being devoted to it.

1.2.6 Summary

This brief outline of some of the industrial requirements for accurate measurement has illustrated: (i) the close links that now exist between measurement capability and the efficiency and reliability of industrial products; and (ii) the existence of new industries and technologies that have grown up simply because certain high accuracy measurements have become possible on a routine basis. Moreover, in some industries, notably the production of microcircuits, commercial viability in a world market worth many billions of dollars, is dependent on fabrication processes at the frontiers of science calling for the most advanced measurement capabilities. In all of these, the pressure to improve measurement capability will continue because economic success may be seen to flow from such improvements. The ability to make accurate measurements will thus continue to be one of the characteristics of successful high-technology industries [9].

1.3 The need for accuracy rather than simply reproducibility

In describing commercial, industrial applications of measurement I have so far used the words 'precision', 'reproducibility' and 'accuracy' without distinguishing them or giving definitions.

The need for what is sometimes called 'absolute accuracy' is often questioned on the grounds that the actual requirement in most practical situations is that measurements be reproducible and uniform from place to place. The difference between reproducible or uniform measurements, on the one hand, and accurate measurements, on the other, is, however, crucial. An accurate measurement is a measurement made in terms of units linked to fundamental physics so that it is (i) repeatable in the long term and (ii) consistent with measurements made in other areas of science or technology. In examining what is done with the most precise measurements it is rarely possible to escape from one or other of these requirements. Thus measurements must be accurate if they are to be really useful and reliable.

A physical or chemical property of a material or substance is quantified by representing it in terms of a unit multiplied by a number. The yield strength of a particular alloy, for example, is a physical quantity whose magnitude may be given by multiplying the unit, in this case the megapascal (MPa), by the number 800 to give 800 MPa; similarly the mass concentration of lead in drinking water may be given as 20 ng m^{-3}. To assign these quantitative descriptions to physical properties or phenomena irrespective of whether or not the highest accuracy is required, suitable units must be defined and ways found to establish the number by which they must be multiplied.

In choosing units, the requirement for long-term stability at the highest level of accuracy leads directly to units linked through physical theory to atomic or quantum phenomena or fundamental physical constants, as these are the only ones whose immutability is sure. This was recognized more than a century ago by Maxwell who wrote in 1870:

> If we wish to obtain standards of length, time, and mass which shall be absolutely permanent, we must seek them not in the dimensions or the motion or the mass of our planet, but in the wavelength, the period of vibration, and the absolute mass of these imperishable unalterable, and perfectly similar molecules.

At that time neither physics nor technology had advanced sufficiently for it to be either feasible or necessary for units to be defined in this way although with the work of Michelson, a few years later, it would have been technically possible to define the metre in terms of the wavelength of light. The existing definition, however, was perfectly satisfactory in practice and there was no strong need from the users to move to a more precise definition until much later.

Consistency of measurements across all areas of science is necessary for the practical application of the equations of physics. For this, the starting point can be obtained only by having a system of quantities and units that is itself consistent. At present, consistency is provided by the Système International d'Unités (SI) with its seven base units, derived units, multiples and submultiples. Among the SI base units, only the kilogram is defined in terms of a material artifact. The second is explicitly defined in terms of an atomic or quantum phenomenon; the metre in terms of the speed of light and the second; the ampere in terms of a mechanical force and a distance; the kelvin in terms of a thermodynamic equilibrium state; the candela in terms of a power and an arbitrary physiological conversion factor; and, finally, the mole is defined in terms of an unspecified number of entities equal to the number of atoms in 12 g of carbon 12. The ensemble of SI base units is linked through physical theory to atomic and quantum phenomena, and to fundamental physical constants through a complex web of experimental measurements. It thus very largely meets the requirements laid out by Maxwell.

When the equations of physics or engineering are used to calculate, for example, the maximum load that an aircraft wing strut will support, it is assumed not only that the results of the calculations are consistent with those done elsewhere or at another time but also that they represent the behaviour of real wing struts under the real foreseen loads. These assumptions are justified only if calculations are based on equations that correctly model physical systems and on data that correctly represent the properties of the alloy from which the strut is made, i.e. on accurate data.

Data that are reproducible but not accurate can misrepresent nature in a multitude of ways. What is sure is that it would be unwise to build aircraft designed on the basis of reproducible but not accurate data as they could not be given any guarantee of safety. Likewise chemical analyses of trace amounts of heavy metals in drinking water that are reproducible but significantly in error can result in damage to human health, and they provide no proper basis for long-term control of water quality.

Without a firm basis in accuracy there is no way of knowing whether apparently reproducible results are constant in time. To the question as to whether the amount of lead in our drinking water is smaller than it was ten years ago, no reliable answer is possible. This leads directly to the subject of the next section.

1.4 Metrology in human health and in environmental protection

I have already discussed the impact of measurement in its role in trade, commerce and the manufacture of high-technology products. In these areas, measurement is a key ingredient and poor measurements lead directly to trade disputes, to inefficient production and to unreliable manufactured products. For many of us, however, measurements have a much more direct influence on our lives when they are part of medical diagnosis or therapy. For better or for worse, medical diagnoses are increasingly made on the basis of the results of measurements. These describe blood pressure, electrical activity of the heart, cholesterol concentration or the levels of many other essential components of the blood. The importance of these measurements is obvious, as is the need for reliability. It is also obvious that accuracies expressed in parts per million are not required. Rather, it is necessary to be sure that errors are not much greater than the smallest detectable physiological effect, usually a few per cent. Without considerable care, errors much greater than this can easily occur. The fact that the accuracies sought are not very high should not disguise the formidable difficulties that must be overcome to achieve these accuracies. Without the efforts that have already been made to assure accuracies of a few per cent in radio-therapy, for example, overdoses or underdoses of a factor of ten (i.e. about one hundred times the present tolerance) would be common. This is because the routine production of well-characterized ionizing radiations is difficult: the radiations themselves are invisible and no immediate physical or biological effects are discernable either to the operator or to the patient.

In addition to its role in medicine, measurement is increasingly important in agriculture, in the testing of food products and in monitoring the environment. Much of this relatively new activity is a consequence of the European Single Market. European Directives (i.e. regulations) on product quality now cover a vast range of products and apply in all countries of the European Union (EU). One of the provisions of the Single Market is that analytical results and tests made by a properly certified and accredited laboratory in one EU country must be accepted in all the others. In preparation for this, the Community Reference Bureau (BCR) has, for a number of years, been carrying on a programme to evaluate the comparability of measurements in different countries and has sought means to improve them. This programme highlighted gross differences in the results of chemical analyses made in different EU countries and was instrumental in bringing about significant improvements.

In many areas, particularly medicine, agriculture and food products, chemical and biochemical analyses were carried out, prior to the BCR work, within closed national systems. Very few cross-border comparisons of these measurements were made because there were no statutory requirements to do so. Astonishing results were obtained when such international comparisons began. One example concerns the determination of aflatoxin B_1 in cattle feed, introduced through peanuts. The permitted levels (EEC Directive 74/63 EEC) are very low, not more than 10 µg kg^{-1} in compound feed, because aflatoxin reappears as a metabolite (aflatoxin M_1) in milk. This aflatoxin is restricted in milk for infants and some EU countries apply limits of 0.01 µg L^{-1}. The early comparisons organized by the BCR show results which differ by a factor of one thousand between countries. Similar differences were found in determinations of trace amounts of heavy metals (such as Cd, Pb, Hg) in milk powder. In each case, the preparation of stable reference materials and the introduction of standardized analytical methods greatly improved the situation. The initial poor results, however, are typical of what can happen when trace amounts of one substance in another are analysed in the absence of a common framework for comparisons.

Despite a superficial simplicity, the problem of making accurate trace analyses is a very difficult one. Difficulties arise because the magnitude of the signal representing the trace substance measured depends on a combination of (i) the real amount of the substance present, (ii) the properties of trace amounts of other substances and (iii) the properties of the substance, known as the matrix, which forms the bulk of the sample. Thus, a real amount of 0.01 µg L^{-1} of aflatoxin M$_1$ in milk could appear on analysis to be quite different depending upon what other trace substances were present and whether the analytical instrument being used had been correctly calibrated against a comparable sample.

Although the problem of achieving accurate trace analysis has been well known for a long time powerful commercial pressures now demand international traceability, and hence accuracy. Their origin is the need to demonstrate conformity with regulations relating to the sale of food and other products. Similar pressure from governments calls for reliability in measurements used as the basis for national legislation and for international agreements dealing with the protection of the environment. In this case it is a question of balancing the economic costs of restricting or changing industrial activity against the environmental penalties of not doing so. The results of measurements are often crucial in coming to a decision.

The importance of improving the accuracy of measurements in analytical chemistry has led the Comité International des Poids et Mesures to establish a new consultative committee to advise it on these matters. International comparisons have already begun at the highest level of accuracy with a view to improving the traceability of such measurements to SI. As was the case for physical and engineering measurements at the end of the 19th century, so now towards the end of the 20th century the requirements of international trade present imperative demands for accuracy and international traceability in measurements in analytical chemistry.

1.5 Metrology and fundamental physics

Confidence in the predictions of physical theory is at the basis of all of today's advanced technology. This confidence exists because the predictions of theory are tested by experiment and theories are rejected, revised or provisionally accepted depending upon the results of the tests. Testing by experiment is a characteristic of modern science which we owe principally to Francis Bacon. It was he who insisted upon the need to carry out real experiments instead of performing the so-called 'thought experiments', which had been used since the time of Aristotle to prove or disprove theories. These 'thought experiments' were no more than descriptions of situations in which the apparently logical consequences of preconceived views of the world were demonstrated. Very little that was new could be discovered, and much that was false was supposedly proved. Bacon showed that it is possible to devise real experiments that can rebut the predictions of theory and so provide a system of choice between one theory and another. The ability to carry out such crucial experiments soon required the ability to make quantitative measurements. In this way the science of measurement became an essential component in the growth of experimental science from its earliest days and it continues to be so today.

1.5.1 Testing of physical theory

The predictions of two of the most fundamental theories of modern physics, quantum electrodynamics (QED) and general relativity (GR), are still being tested to establish the limits of their applicability and, in the case of general relativity at least, to distinguish them from those of competing theories. By their very nature, the testing of the predictions of both of these theories requires careful and difficult measurements. In the case of QED this is because it deals with the microscopic world of atoms and nuclei, and in the case of GR because in the local universe, the only one to which we have easy access, the differences between the predictions of GR and those of Newtonian physics are extremely small.

To test QED, one experiment consists of making measurements of the anomalous magnetic moment of the electron so that the results can be compared with the value that is predicted by calculation directly from QED. At present the value obtained from the best measurements differs by less than its experimental uncertainty, a few parts in 10^9, from the calculated value.

To test GR is more difficult. Until the beginning of space flight in the 1970s, GR was a domain of theoretical physics almost wholly disconnected from the rest of physics and with no immediate practical application. This has all changed. Any yachtsman who uses a GPS receiver to fix a position at sea depends on corrections for the curvature of spacetime that are applied to the rates of the atomic clocks on board the GPS satellites, in orbit 20 000 km above. With the accuracy of present-day commercial atomic clocks, a few parts in 10^{13} per day, the equations of GR are relatively straightforward to apply. This will not be the case when the next generation of atomic clocks, having accuracies two or three orders of magnitude better than the present ones, becomes available. Who would have thought only 20 years ago that GR would become a routine part of space engineering and that equipment on sale to the general public would make use of satellite signals corrected for the curvature of spacetime!

On a much more down to earth level, discoveries in solid state physics such as the Josephson or quantum-Hall effects make startling and simple predictions concerning the properties of semiconductors at low temperatures. That the Josephson voltage is exactly proportional to integral steps of $2e/h$ or that the quantum-Hall resistance is exactly proportional to integral steps of h/e^2 (where e is the charge on the electron and h is Planck's constant) merited the discoverers of these effects their Nobel Prizes, but their predictions have been subject to the most exacting measurements to test whether or not they are universally valid. These and similar discoveries in solid state physics are part of the essential basis of the science upon which today's micro-electronics industries are built.

1.5.2 The fundamental physical constants

So far great emphasis has been placed on the importance of measurement standards being linked through the theories of physics to atomic or quantum phenomena, or to fundamental physical constants, so that they can be assured of long-term stability. What are these so-called fundamental physical constants? The theories of science point to the existence of constants of nature and to certain invariant relations that exist between them. These constants appear quite independently in different areas of science. The fact that they take the same value, no matter what type of phenomenon is being considered, indicates that they are of a very fundamental nature and hence their name 'the fundamental physical constants'.

Some of these constants appear directly from theory. These include the speed of light in Einstein's Special Theory of Relativity, the Planck constant in quantum theory and the Sommerfeld fine-structure constant in QED. Others are properties of the elementary particles, such as the mass and charge of the electron or the mass and magnetic moment of the proton. In a third category are those constants that are, in effect, conversion factors between different types of quantities. Among these are the Boltzmann constant, which is the link between thermal and mechanical energy, and the Avogadro constant, which relates microscopic to macroscopic amounts of matter.

The values of the fundamental constants are obtained in SI units either directly from experiments, as is the case of the magnetic moment of the proton and the Boltzmann constant, or indirectly by calculation, using one of the many relationships that exist between the constants. The universal nature of these constants allows relationships to be found between them that span wide areas of science and are susceptible to experimental test in diverse ways. Indeed, to obtain good agreement between the measured values of these constants obtained using methods based upon quite different areas of physics is one of the ways of confirming the consistency of physical theory.

Accurate measurements of the fundamental physical constants contribute to advances in physics and to metrology and are essential, in the long term, to both [10, 11]. The most recent example of the symbiotic relation between metrology physics and high-technology industry is to be found in the quantum-Hall effect discovered in 1980. As a practical standard of electrical resistance, the quantum-Hall effect is widely used in national metrology laboratories and even in some industrial companies. The technology developed to use the quantum-Hall effect for metrology, based upon that used in commercial microcircuits, has at the same time been used in accurate experiments to study the physics of a two-dimensional electron gas in GaAs heterostructures. This has turned out to be a very complex phenomenon and much remains to be understood, but it has opened up new avenues of research in physics which, in their turn, will lead to new applications in semiconductor electronics.

1.6 The role of national and regional metrology organizations

Practical application of the SI, for the overwhelming majority of users of measuring equipment who do not themselves have the means for the direct realization of the SI units, is through a national measurement system based upon units maintained or realized in a national metrology laboratory.

1.6.1 The origins of national metrology laboratories

Towards the end of the 19th century, the needs of industry for accurate measurement standards and of international trade for worldwide agreement of measurement standards led to the Convention du Mètre and the foundation of the major national standards laboratories. The objectives of these first national standard laboratories were quite clear. They were to support national manufacturing industry, to establish national measurement standards, to provide calibrations and to ensure comparability with the national standards of other countries. In those days there existed, for almost all measurement standards, a clear hierarchical chain extending from the national standard to the workshop bench. Traceability, in the sense of a continuous chain of calibration certificates, soon extended throughout individual nations and then across the world through international comparisons of national standards.

From the earliest days of the national laboratories until the early 1970s, most high-level calibrations for industrial clients were carried out in the national laboratories themselves. This is no longer the case. The number of such calibrations has outgrown the capacity of national laboratories and most calibrations for industry are now carried out in national networks of certified or accredited calibration laboratories. This change has taken place for two reasons; first, it reflects an increasing demand for formal traceability of measurements and second, it is part of a worldwide move to establish formal national measurement systems widely dispersed in the community. The national standards or metrology laboratory remains, however, the source of the measurement standards and expertise which support the calibration services. Without them, these services could not function. For example, in 1992 the PTB carried out a total of 671 calibrations for the laboratories of the German calibration service DKD. These in turn carried out more than 24 000 calibrations for industrial concerns which were then used as the basis of more than 1.3 million internal calibrations in just three of the largest industrial users of accurate measurements. Similar figures could, of course, be given for other countries and they illustrate the key role played by national standards laboratories in the support of high-technology industrial activity of a country [12]. The national laboratories also, of course, compare their national standards with those of other countries. This is done either through the BIPM or through local regional metrology organizations [13].

1.6.2 The need for a research base in measurement science

Measurement standards are not static. They evolve continually to reflect advances in physics and in response to changing industrial needs so national laboratories must maintain an active research base in measurement science. This is necessary so that national industries can obtain the most advanced and accurate calibrations in their own country. Research in measurement science is a long-term activity that must necessarily be done in advance of industrial and other requirements. Today's research in metrology provides the basis for tomorrow's calibration services.

The national benefits of an active research base in a metrology laboratory are not only long term, they are also immediately available through the expertise that results uniquely from an engagement in active research. Major national metrology laboratories have thousands of industrial visitors each year; they run courses and seminars, and are represented on all the important industrial standards bodies. These close contacts with national industries provide important feedback to the national laboratory on present and future industrial measurement requirements.

The long-term research also provides a fund of knowledge and expertise that frequently leads to unexpected applications of the results to pressing practical problems. A recent example of this is the application at the NIST of silicon crystal x-ray diffraction spectrometry to diagnostic x-ray mammography. The image quality in all soft tissue x-rays is critically dependent on the spectrum of the x-rays, which in turn is controlled by the high voltage (20–30 kV) applied to the x-ray source. It is difficult to measure these high voltages with sufficient accuracy under the conditions required for routine diagnostics. An elegant solution has been found from the long-term programme of basic metrological research at the NIST related to the measurement of the crystal lattice spacing of silicon using x-ray diffraction with a view to the development of a new standard of mass based on the mass of an atom of silicon. This work has not yet resulted in a new standard of mass but it has, among other things, led to a great deal of knowledge on x-ray spectroscopy of silicon crystals. A silicon crystal x-ray diffraction spectrometer, developed in this work, appears to be an almost ideal way of routinely measuring x-ray spectra from x-ray diagnostic machines.

The advantages to be gained by having all measurement standards in one institute have long been recognized. The close links that now exist between all areas of advanced metrology through their dependence on basic physics reinforce this view. The fact that research on a wide range of measurement standards is carried out in one location has, in addition to the obvious advantages of scale in terms of common services, the great advantage of concentrating intellectual resources where they can be most easily tapped either from within the laboratory or from outside. The application of silicon crystal x-ray spectrometry to x-ray mammography is in fact a direct result of a request to the NIST from the US Conference of Radiation Control Program Directors for help, which was then circulated within the NIST and came to the notice of those responsible for the silicon crystal research.

A national institution devoted to metrology should also become well known throughout the country, should provide easy access for those needing advice and be clear point of reference for all questions related to measurement. A recent study [14] of what industrial companies expect to gain from contacts with Federal laboratories in the US, for example, showed that easy and direct access to specialized technical resources was voted an equal first in importance with the possibility of collaborative projects. Great emphasis was placed on the need for such laboratories to make widely known the range and depth of their activities and to facilitate outside access. The conclusion of this study should be taken careful note of by national metrology laboratories whose technical resources are often unique in their country.

1.6.3   International representation

Global industrial and commercial activity is increasingly regulated at a technical level through international standards (ISO, IEC) and through international agreements such as those made under the International Radio Consultative Committee (CCIR) or the Convention du Mètre. These agreements and standards are drawn up by international committees whose members are experts nominated by their governments. The scientific or technical competence of the individual expert strongly influences the decisions made by such committees. National interests, which may have important commercial or prestige aspects, thus depend on having strong representation on these committees. In metrology, national representatives on international committees are drawn almost exclusively from national metrology laboratories. It is thus in the national interest that the national laboratory, irrespective of its size, provides good representation for international committees.

1.6.4   International cooperation between metrology laboratories

In the past, international cooperation in metrology was carried out almost exclusively through the BIPM and concerned only national standards at the highest level. This is no longer the case. The development of national calibration services, the certification and accreditation of calibration and testing laboratories has led to international links between these services. It has also led to greater activity in the international comparison of standards at levels other than the national primary ones.

The increase in international activity in metrology at levels other than that dealt with by the BIPM, began in the 1970s with the creation by the Commission of the European Communities of its Community Reference Bureau (BCR), recently renamed the Measurement and Testing Programme. The BCR had, as part of its programme, the support of applied metrology in the EC countries. The major part of its early effort, however, was the establishment of traceability and good measurement practice in the vast range of chemical analyses related to agriculture, food and, to a lesser extent, medicine. The success of the BCR in showing up, and in due course reducing, some of the very wide divergences that existed between analytical laboratories in different countries of the then European Community has already been remarked upon in section 1.4. In 1987 stimulated by the activities of the BCR and the growing level of European collaboration in other areas, the national laboratories of the EC countries, together with a number of those from the EFTA (European Free Trade Association) countries, created EUROMET. EUROMET is a European collaboration in measurement standards and in 1993 has 18 participating national metrology institutes plus the Commission of the EC. The stated aims of EUROMET are:

- to develop a closer collaboration between members in the work on measurement standards within the present decentralized metrological structure;
- to optimize the utilization of resources and services of members and emphasize the deployment of these towards perceived metrological needs;
- to improve measurement services and make them accessible to all members; and
- to ensure that national facilities developed in the context of EUROMET collaboration are accessible to all members.

Similar regional metrology organizations have since been set up in other parts of the world, notably the Asia/Pacific Metrology Programme, which includes members from the Asia/Pacific region including Australia and China, and NORAMET which comprises the national metrology laboratories of Canada, Mexico and the United States‖. Similar regional metrology organizations are also being formed in South America and Eastern Europe. At present the most active of these regional organizations is EUROMET. All this regional activity is complementary to that of the BIPM which ensures worldwide traceability of measurement standards. The BIPM now takes care to include, as far as possible, at least one representative from each of the regional organizations in its own international comparisons of measurement standards. For metrology in chemistry, the work of the BCR also stimulated the European laboratories responsible for chemical analysis to form EURACHEM which is to the field of analytical chemistry what EUROMET is to the field of physical measurement. Founded in 1989 by a group of directors of laboratories from ten countries in Europe, EURACHEM now has members from 17 countries and includes delegates from governments, universities and industry. The aim of EURACHEM is to provide a framework in which analysts can collaborate and improve the accuracy of chemical measurements.

‖ NORAMET is now one of the five regions of the Interamerican System of Metrology (SIM). The other regions are CARIMET (Caribbean countries), CAMET (Central American countries), ANDIMET (Bolivia, Colombia, Ecuador, Perú and Venezuela) and SURAMET (Argentina, Brazil, Chile, Paraguay and Uruguay).

1.7 The cost of maintaining a national measurement system

Measurement and measurement-related operations have been estimated to account for between 3% and 6% of the GDP of industrialized countries [15, 16].

The cost of maintaining a national measurement system in an industrialized country is between about 30 and 70 parts in 10^6 of the GDP. For the 12 EU countries the total direct spending on national measurement systems in 1992 was about 400 Mecu, equivalent to nearly 70 parts in 10^6 of the GDP of the Community. For the UK and Germany the figures are similar, but a little less than this. For the USA the cost of metrology and related activities of the NIST alone, 170 million dollars, represents 30 parts in 10^6 of the 1992 GDP and for Japan the figure is between 20 and 40 parts in 10^6 of GDP depending upon whether or not the industrial metrology at the Prefecture level is taken into account. Not included in the figure of 30 parts in 10^6 for the USA, however, are other costs related to the national measurement system that are included in the EU figures. In the USA, for example, the NCSL (National Conference of Standards Laboratories, some 1200 members) and the Weights and Measures Laboratories (members from each of the States) are integral parts of the national measurement system and are not funded through NIST. The announced intention of the US government to double the direct funding of NIST within three years should be noted. There should, nevertheless, be economies of scale in the USA and Japan which have central national metrology institutes, relative to the EU in which each member nation has its own metrology institute. In some rapidly developing countries in the Asia/Pacific region, expenditure on establishing a national measurement system reaches 100 parts in 10^6 of GDP. In terms of the figures given here, the cost of maintaining a national measurement system represents a few tenths of a per cent of the 3–6% estimated to be spent by industrialized nations on measurement and measurement-related operations.

At an international level, the principal cost of maintaining the international measurement system is that of supporting the BIPM, which, in 1992, was seven million dollars. This represents about 0.4 parts in 10^6 of the GDP of each of the member nations of the Convention du Mètre and is, on average, less than 1% of what each country spends on its own national metrology laboratory.

1.8 Final remarks—a personal view

In concluding this overview of the field of measurement science and its role in society, I would like to highlight just a few of what I consider to be the most important features:

- Metrology is an essential but largely hidden part of the infrastructure of today's world.

- The economic success of most manufacturing industries is critically dependent on how well products are made and measurement plays a key role in this.
- Human health and safety depend on reliable measurements in diagnosis and treatment.
- The protection of the environment from the short- and long-term destructive effects of industrial activity can only be assured on the basis of reliable and, therefore, accurate measurements.
- Physical theory, upon which all of today's high technology and tomorrow's developments are based, is reliable only to the extent that its predictions can be verified quantitatively.
- There exists a world-wide measurement system maintained under an intergovernmental convention which assures the uniformity and accuracy of today's measurement standards and provides the research base in measurement science that will allow tomorrow's standards to be developed.
- Between 3% and 6% of GPD in industrialized countries is devoted to measurement and measurement-related activities. The cost of providing a national measurement system is a few tenths of a per cent of this, equivalent to between 30 and 70 parts in 10^6 of GDP. A further 0.4 parts in 10^6 of GDP goes to providing support for the international measurement system through the BIPM.

At the beginning of this chapter I referred to the difficulties that sometimes arise in justifying to governments the costs of national measurement systems—principally for the maintenance of national metrology laboratories. In describing the activities of national metrology laboratories I laid great stress on the benefits that an active metrology laboratory brings to national affairs through its support of industry and the provision of advice to government and national representation on international bodies.

I end by urging governments to maintain their metrology laboratories and most particularly to maintain their active research base. Without long-term research it is not possible to meet industrial requirements for accurate measurement standards and the research base, once lost, can only be re-established at great cost and over a considerable time. The national metrology laboratories are national assets whose influence is largely hidden and diffuse, but represent a significant and cost-effective contribution to national affairs.

References to chapter 1

[1] *The International System of Units (SI)* 1991, 6th edn (BIPM)
[2] Swyt D A 1992 *NIST Report* 4757. Challenges to NIST in dimensional metrology: the impact of tightening tolerances in the US discrete-part manufacturing industry
[3] McKeown P A 1992 High precision manufacturing in an advanced industrial economy *Metrology at the Frontiers of Physics and Technology (Proc. CX Int.*

School of Physics 'Enrico Fermi') ed L Crovini and T J Quinn (Amsterdam: North-Holland) pp 605–46
[4] Kunzmann H 1992 Nanometrology at the PTB *Metrologia* **28** 443–53
[5] Yoshida S 1992 Nanometrology *Metrologia* **28** 433–41
[6] Leach R, Haycocks J, Jackson K, Lewis A, Oldfield S and Yacoot A 2000 Advances in traceable nanometrology at the NPL *Proc. Int. Seminar on the Future Directions for Nanotechnology in Europe and Japan (Warwick University, 18–19 September 2000)* pp 112–28 (not in the original paper by Dr Quinn)
[7] Kartaschoff P 1991 Synchronization in digital communications networks *IEEE Instrum. Meas.* **79** 1019–28
[8] Yuan L, Authes R A, Ware R H, Rocken C, Bonner W D, Bevis M G and Businger S 1993 Sensing climate change using the global positioning system *J. Geophys. Res.* **98** 14 925–37
[9] White R M 1993 Competition measures *IEEE Spectrum* **April** 29–33
[10] Cohen E R and Taylor B N 1986 The 1986 adjustment of the fundamental physical constants *Codata Bulletin* no 63 (Oxford: Pergamon)
[11] 1983 Research concerning metrology and fundamental constants *Report of the Committee on Fundamental Constants* (Washington, DC: National Research Council, National Academy Press)
[12] Kose V, Brinkmann K and Fay E 1993 15 Jahre Deutscher Kalibrierdienst *PTB Mitteilungen* **103** 51–60
[13] Braun E and Kind D 1992 Metrology in electricity *Abhandlungen Braunsch. Wiss. Gesellschaft* **43** 159–75
[14] Roessner D J 1993 What companies want from the federal laboratories *Issues in Science and Technology* **10** 37–42
[15] Don Vito P A 1984 Estimates of the cost of measurement in the US economy *Planning Report* 21 (NBS: US Department of Commerce)
[16] Clapham P B 1992 Measurement for what is worth *Eng. Sci. Ed. J.* **August** 173–9

Chapter 2

Basic concepts and definitions

The ever increasing globalization of national economies has had important consequences in the areas of metrology, standardization, testing and quality assurance. One of these consequences is the widespread recognition of the relevance of measurements to make decisions about the quality of products and services. It is recognized as well that the information about quantities that results from their measurement can rarely be assumed to be complete. For this reason, confidence in measurements is possible only if a quantitative and reliable expression of their relative quality, the measurement uncertainty, is assessed.

The concept of uncertainty will be introduced in this chapter rather intuitively using a simple example taken from the field of *conformity assessment*. In general, this includes all activities concerned with determining that relevant requirements in standards or regulations are fulfilled. Conformity assessment procedures provide a means of ensuring that products, services or systems produced or operated have the required characteristics, and that these characteristics are consistent from product to product, service to service or system to system. The outcome of the assessment is a decision to accept or reject the inspected item or lot; and this decision is often in turn based on the results of measurements performed. In order not to make the wrong decision, an appraisal of the measurement uncertainty is crucial.

In discussing the example, it will become apparent that the act of measurement has to be properly defined and understood. The definition includes concepts such as quantities, values and units of measurement, all of which will be defined and explained as well. Other essential concepts, those of calibration and traceability, will appear soon afterwards.

Uncertainty and measurement error are often mistakenly considered as synonyms. A similar confusion involves the concepts of accuracy and precision. A brief discussion about the distinction between these pairs of terms will be given.

Finally, the concept of measurement model will be defined and illustrated. As will be made clear in chapter 3, a proper model formulation is of paramount importance for the evaluation of uncertainty.

26   Basic concepts and definitions

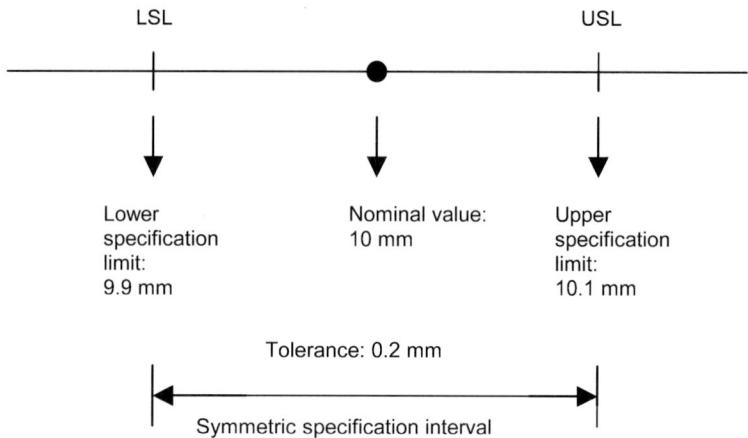

Figure 2.1. A symmetric specification interval.

2.1 An example: tolerance checking

Measurements are usually needed to quantify product characteristics or to control the variables of production processes in order to make sure that these characteristics or variables stay within prescribed limits. Many times, such measurement results are compared with the *specifications* that have been set forth for the products or processes to be acceptable. In most cases, a specification consists of a desired *nominal value* together with its *tolerance*. Both values serve to define an interval of allowed variability, such that if the measurement result falls within this interval, the measured item is considered to be acceptable.

For example, suppose the specification for the length of metal rods to be used as spacers is 10 mm for the nominal value and 0.2 mm for the tolerance. If the tolerance is symmetric with respect to the nominal value, this specification may be written conveniently as (10 ± 0.1) mm. This means that, to be accepted, a given rod should measure between 9.9 mm (the lower specification limit, LSL) and 10.1 mm (the upper specification limit, USL). This is represented graphically in figure 2.1. (In general, however, the specification interval does not need to be symmetric.)

The company's quality control inspector has to measure each individual item to determine whether it should be accepted or rejected. Suppose he/she picks a rod from the production lot, measures it with a caliper and obtains the value 9.88 mm. Being very conscientious, the inspector takes a second measurement; this time he/she obtains 9.91 mm (figure 2.2). The first measurement would lead to rejection, the second to acceptance. Is there something wrong with the measurement procedure? Should only one value be obtained every time the rod is measured?

An example: tolerance checking

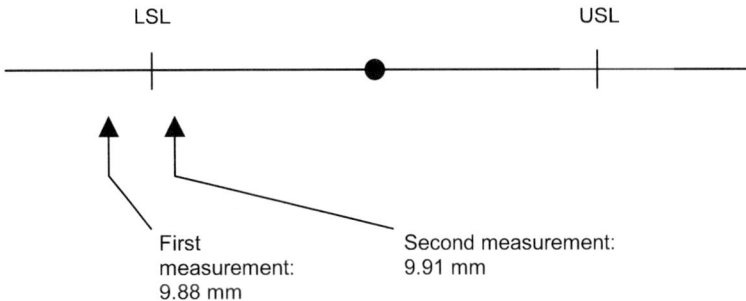

Figure 2.2. Two possible measurement results.

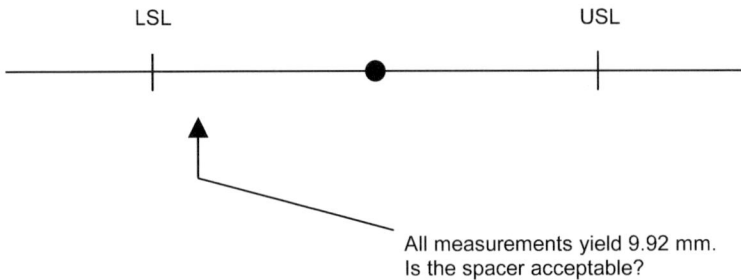

Figure 2.3. A series of repeated measurements yields the same value.

The answer to both these questions is 'not necessarily'. In effect, the fact that different results were obtained in the two apparently identical measurement operations should not be surprising. For each time the inspector may have measured the piece in slightly different positions, a different pressure may have been applied on the caliper, the temperature of the spacer may have been altered by heat transfer from the hand of the inspector, and so forth. These types of effects, which affect the measurement results and cannot in general be eliminated nor precisely controlled, are called *influence quantities*.

Let us simplify matters for the moment and assume that the action of the influence quantities is, in this case, negligible, such that several repeated measurements yield the same result, say 9.92 mm (figure 2.3). Can the inspector now be sure that the spacer abides by its length specification? The answer is again 'not necessarily', because the length of the rod may not be *actually* 9.92 mm. To explain this apparent contradiction the meaning of measurement needs to explored.

28   Basic concepts and definitions

2.2   Values and units of measurement

According to the ISO *International Vocabulary of Basic and General Terms in Metrology* [20]†, popularly known as the VIM, the act of *measurement* is defined as the set of operations having the objective of determining a value of a *quantity*. The latter is an attribute of a phenomenon, body or substance that may be distinguished qualitatively and determined quantitatively. Quantities can be interpreted in a *general* sense, e.g. length, or may refer to *particular attributes*, such as the length of our rod. Naturally, it only makes sense to measure a given particular quantity, called the *measurand*, not quantities in the general sense.

In turn, the word *value* refers to a quantitative algebraic expression involving the product of a number (referred to as the *numerical value*) and the *unit of measurement*, or simply *unit*:

$$\text{Value} = \text{Numerical value} \times \text{Unit}.$$

Measurement results are usually given in this way, but with an implicit multiplication sign (adimensional quantities, for which there are no units, are an exception. In such cases the value and the numerical value become equal to each other. See [39] for a discussion on adimensional quantities).

The unit is a particular quantity, adopted and defined by convention, with which other quantities of the same kind are compared in order to express their magnitudes relative to that quantity. We may then redefine the act of measurement as the set of operations by which the numerical value, i.e. the ratio between the value of the measurand and the value of the unit, is established. Thus, from the previous equation

$$\text{Numerical value} = \frac{\text{Value (of the measurand)}}{\text{(Value of the) unit}}.$$

The value of the unit is represented by a conventional sign known as the unit's *symbol*. In our example the result 9.92 mm means that the length of the spacer was determined to be 9.92 as large as 1 millimetre (symbol mm). The latter is in turn defined to be exactly equal to a length of 1 metre (symbol m) divided by 1000. Now, what is a metre? Or more properly, which is the particular length that has been adopted and defined by convention to be exactly equal to one metre? The answer is 'the length of the path travelled by light in vacuum during a time interval of 1/299 792 458 of a second' (figure 2.4). Thus, in order for the inspector's result to agree with this definition, it should be possible to assert that light takes 9.92/299 792 458 000 seconds to travel in vacuum a distance defined by the two ends of the rod!‡

† References in chapters 2–6 are listed alphabetically starting on page 236.
‡ Information on the SI system of units, from which the definition of the metre here was taken, can be found on a number of web pages, e.g. www.bipm.fr/enus/3_SI/si.html. This book observes the rules and style conventions of this system.

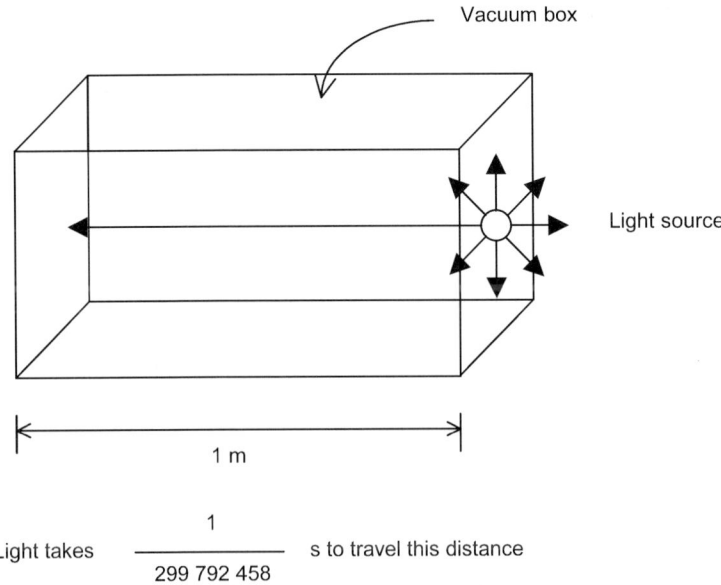

Figure 2.4. The definition of the metre.

Of course, it is not practical to determine the rod's length by means of a vacuum flask, a light source and a stopwatch. Instead, the inspector uses a caliper, which is a particular instance of a device known in general as a *measuring instrument* or simply *instrument*. In daily life we normally take for granted that the indications of the instruments we constantly use give results in units that agree with their definition. In other words, it is seldom the case we would call those results into question. Examples are the volume of gasoline we put in the car, the monthly electric power consumption at home or the weight of the potatoes we buy at the local market. This is because the degree of accuracy required in such cases is not very high. For example, if instead of one kilogram of potatoes we actually got 1010 g, most of us would not get very happy, nor very sad (or mad) if we got 990 g.

However, unwarranted assumptions are not acceptable in industry. In our example, as little as 0.01 mm difference between the 'real' and the indicated values the inspector obtains might signify the rejection of an acceptable item or *vice versa*. For this reason, an essential requirement to ensure the quality of measurements is that measuring instruments be *traceable*. The VIM's definition of this concept will be given in section 2.3, for the moment it can be taken to mean that the instrument must be periodically tested against a measurement standard whose value is known and agrees with the definition of the measuring unit it realizes. The results of that process, known as *calibration*, consist of the values by which the instrument's indications must be corrected in order for them to agree with the definition of the unit.

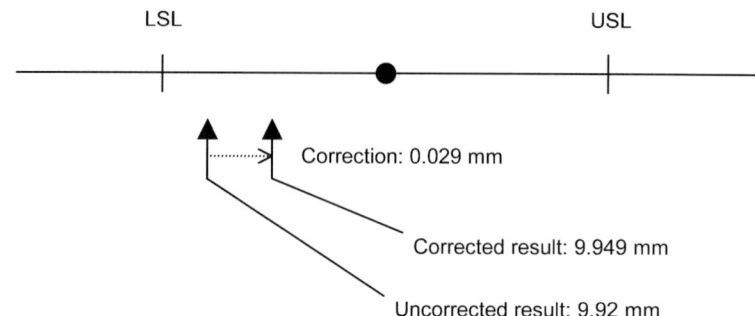

Figure 2.5. The corrected measurement result.

For example, suppose that the caliper used by the quality inspector has been calibrated at 5 mm intervals. The calibration certificate establishes that the *correction* at 10 mm is +0.029 mm. What this means is that to indications in the vicinity of the scale at 10 mm the value 0.029 mm has to be added in order for the *corrected result* to agree with the definition of the unit. In our case all uncorrected indications were 9.92 mm, hence the corrected result is (9.92 + 0.029) mm = 9.949 mm (figure 2.5). As this value is well within the spacer's specification interval, and it agrees with the unit's definition, should the inspector finally accept the rod as conforming with its specification? The answer is 'not yet', because he/she needs to consider one more element of the measurement result. However, before addressing this point, it is convenient to understand how correction values are established.

2.3 Calibration and traceability

The VIM defines calibration as

> the set of operations that establishes, under specified conditions, the relationship between the value of a quantity indicated by a measuring instrument and the corresponding value realized by a measurement standard§.

In general, a *measurement standard*, or simply a *standard*, is a material measure, measuring instrument, reference material or measuring system intended to define, realize, conserve or reproduce a unit or one or more values of a quantity to serve as a reference.

Calipers are normally calibrated using standards of the *material measure* type. These are devices intended to reproduce or supply, in a permanent manner during their use, one or more known values of a given quantity. Examples include

§ This is a simplified definition. The complete definition of calibration is given in section 3.11.

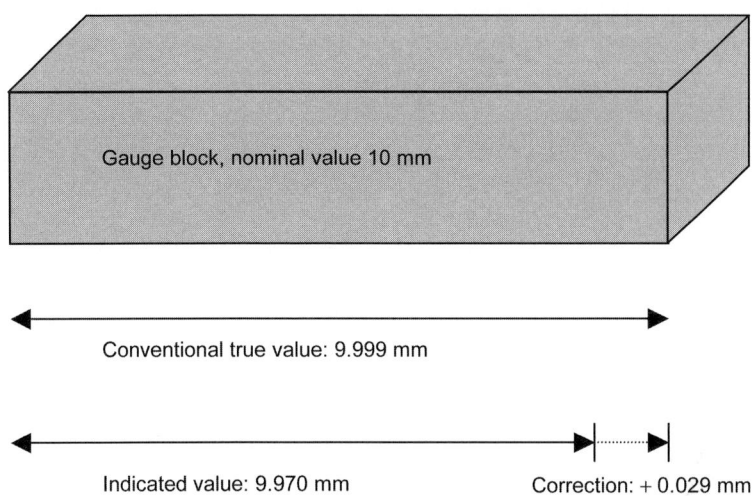

Figure 2.6. Calibration using a gauge block.

certain types of thermometers, sets of masses, electrical resistors, high-purity reference materials, etc. In length, measurements standards are usually metal or ceramic box-shaped solid artifacts known as *gauge blocks*. The quantity realized by a gauge block is the distance between two of its parallel (measuring) faces. Gauge blocks usually come as a set, each block having a specified *nominal value*, that is, a rounded or approximate value of the block's measuring length that provides a guide to its use.

To calibrate a caliper at its 10 mm scale indication a block of that nominal length is selected. The *actual* length of the block is not known. However, its *conventional true value* (a concept to be defined in section 2.5) needs to have been established. Assume that this value is 9.999 mm. During calibration the block is 'measured' with the caliper, and the value 9.970 mm is obtained (normally, as the average of many indications). The correction is therefore (9.999 − 9.970) mm = 0.029 mm (figure 2.6). Thus, a calibration may be redefined as a measurement operation where the measurand is a correction to the value indicated by the calibrand.

The conventional true value of a standard must be determined through its calibration with another standard. For example, the length 9.999 mm may have been determined by means of a mechanical comparator that measured −0.002 mm for the difference between the length of the block and another block of the same nominal length and conventional true value equal to 10.001 mm. The latter length in turn may have been determined by another calibration, possibly using a stabilized laser interferometer (another standard), whose indications can be linked via time and frequency considerations to the definition of the metre given earlier.

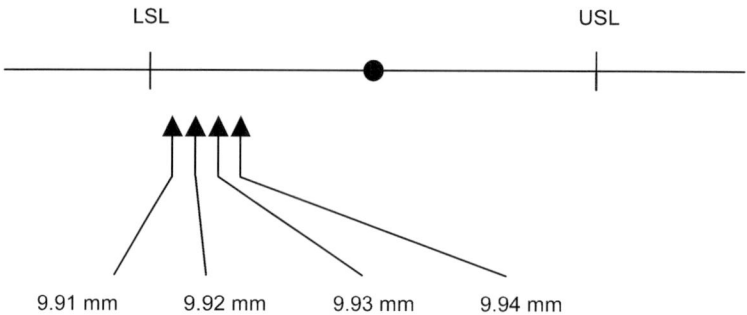

Figure 2.7. Resolution: the caliper cannot distinguish between uncorrected lengths in between those shown here.

We now come to the (partial) definition of traceability as

the property of the result of a measurement whereby it can be related to the unit of measurement through an unbroken chain of comparisons...

A traceable *corrected* result, such as 9.949 mm for the length of the spacer, assures its *reproducibility*: the same result, or one very close to it, should be obtained if the spacer was to be measured by another inspector, at a different place and possibly using another (calibrated) instrument. The quality implications of the traceability requirement should now be clear.

2.4 Measurement uncertainty

It was mentioned in section 2.2 that while the 9.949 mm result for the length of the rod is within the specification interval for this product, it is not clear yet whether the spacer should be accepted. This is because of the uncertainty that needs to be attached to the measurement result. Without being very precise yet about the meaning of the term 'uncertainty', let us assume that *twice its value* represents a region of about 9.949 mm within which the *actual* (and unknown) value of the rod is expected to lie. Why should such a region exist?

Consider first an important characteristic of most measuring instruments, their *resolution*. This is defined in the VIM as 'the smallest difference between indications that can be meaningfully distinguished'. Suppose that the resolution of the caliper is 0.01 mm. This means that the instrument can only give (uncorrected) indications such as 9.91 mm, 9.92 mm and so forth (figure 2.7). If an instrument with a higher resolution was used, giving (uncorrected) indications between 9.915 mm and 9.925 mm, the caliper would give instead 9.92 mm. Hence, considering the 0.029 mm value for the correction, the spacer's length could be anywhere between 9.944 mm and 9.954 mm (figure 2.8).

Another source of uncertainty has to do with the measurand's *definition*. Defining a measurand to be 'the' length of a given rod implies that its end faces

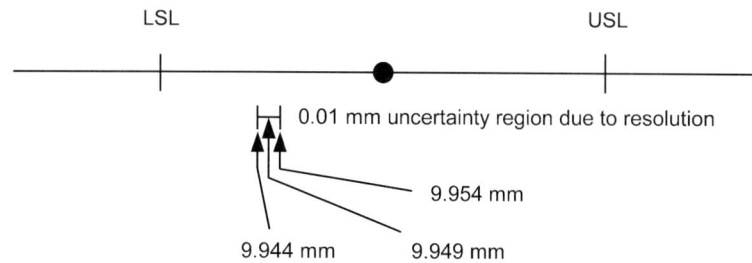

Figure 2.8. Uncertainty due to resolution.

are assumed to be perfectly flat and parallel. Any deviations in these attributes would produce an *incomplete* definition of the measurand, because the actual length that corresponds to a given measurement would depend on the point where the instrument is placed. However, even if the measurand were to be perfectly defined in the geometric sense, one would still need to specify at what temperature the length was to be measured. But then again, could the temperature be measured and maintained at the specified value? In fact, temperature is a very common influence quantity.

The term *repeatability* has been coined to express the closeness of the agreement between the results of successive measurements of the same measurand carried out by the same observer, using the same instrument and measurement procedure under the same conditions, in the same location and with repetitions over a short period of time. These are called *repeatability conditions* and describe a situation whereby every effort is made to minimize the effects of influence quantities. However, since these influences cannot be altogether excluded, repeatable measurements will yield, in general, variable results. Therefore, one cannot ascertain that any one of those results is equal to the 'correct' value. With none of them being 'better' than the others, their average is the most 'honest' way of *estimating*—not asserting—the uncorrected value of the measurand.

Consider now the corrections to be added to the uncorrected result. One of these corrections is the calibration correction, but there may be others. Suppose the measurand is defined to be the length of the rod at a specified temperature T_o. For some reason it is only possible to carry out the measurement at another temperature $T \neq T_o$, and then not with a caliper, but with an instrument that is not affected by temperature. The uncorrected result would then need to be multiplied by the correction factor $F = [1 + \alpha(T - T_o)]^{-1}$, where α is the linear coefficient of thermal expansion of the rod's material. Now T and α have to be measured, each with their own uncertainty. These uncertainties contribute to the uncertainty in F, which in turn contributes to the uncertainty in our estimate of the value of the rod at T_o.

Calibration corrections, while their estimated values may negligibly affect the uncorrected result, also have associated uncertainties. These uncertainties are

34   Basic concepts and definitions

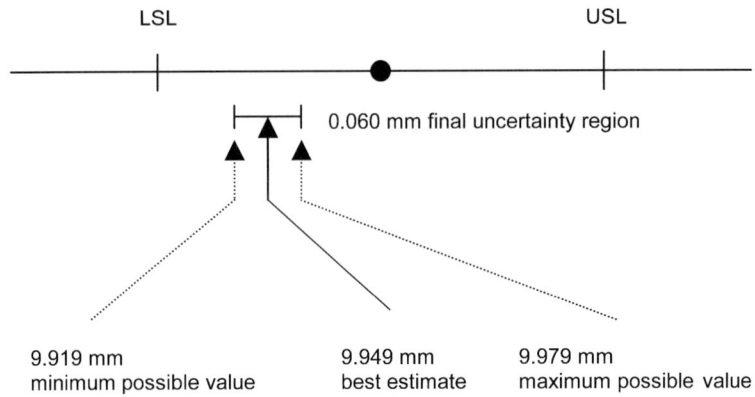

Figure 2.9. Final best estimate and associated uncertainty region.

a combination of the uncertainty in the standard's value and the uncertainty that arises in the calibration process. Hence, the uncertainty increases as one proceeds down from the definition of the unit in the traceability chain. To determine the uncertainty in the final calibration correction all intermediate uncertainties have to have been established. For this reason, the definition of traceability on page 32 ends with the final sentence: '... an unbroken chain of comparisons, *all having stated uncertainties*'.

To continue with our example, let us assume that, after considering carefully all factors that determine the uncertainty in the rod's length, it has been determined that it cannot be less than 9.919 mm nor more than 9.979 mm, with a best estimate equal to 9.949 mm (that is, the uncertainty has been determined to be equal to 0.030 mm). Since the minimum and maximum possible lengths are respectively greater and smaller than the lower and upper specification limits (9.9 mm and 10.1 mm, respectively), the inspector can finally be fully confident in accepting the measured spacer (figure 2.9).

Consider now what would happen if the uncertainty was larger. Suppose that the best estimated value is again 9.949 mm but this time the uncertainty is, say, 0.060 mm. Then, the minimum possible value, 9.889 mm, is below the lower specification limit (figure 2.10). It is, of course, very *probable* that the actual value conforms to the rod's specification. However the inspector cannot be sure about this, given the information he/she has. Therefore, if he/she does not want to run the risk of accepting a non-complying spacer, he/she should reject the item.

This conclusion can be generalized as follows. Let S be the width of the specification interval (in appropriate units) and let U be the measurement uncertainty. Only those measurements that fall within an acceptance interval A of width $S - 2U$ can be accepted safely (figure 2.11).

In reality, things are a little more complicated. In fact, a measurement result expressed as, say, (9.949 ± 0.030) mm should be interpreted as meaning that

Measurement error, accuracy and precision

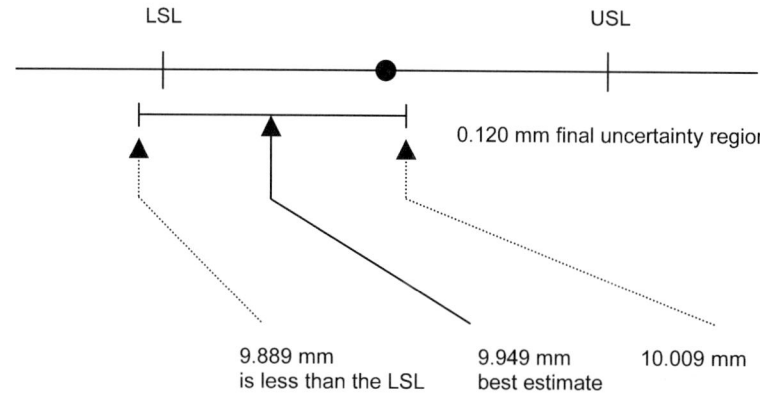

Figure 2.10. Final best estimate with a larger uncertainty.

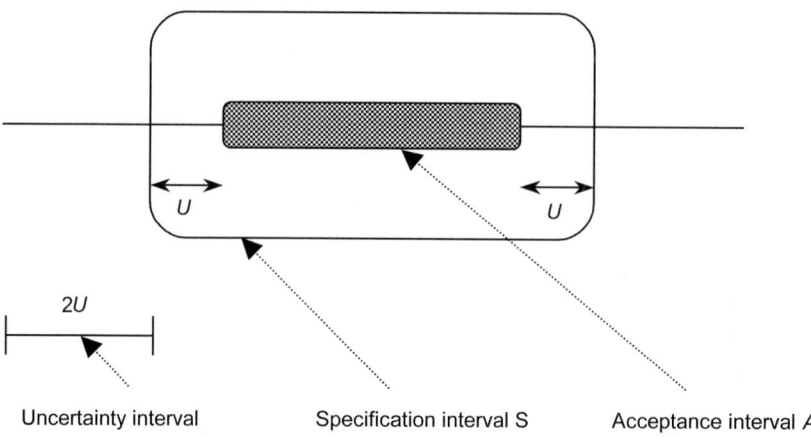

Figure 2.11. The width of the acceptance interval is reduced because of the measurement uncertainty.

the measurand is *very probably* not greater than 9.979 mm nor smaller than 9.919 mm. The probability that the actual value of the measurand falls outside these limits could be made to approach zero by *expanding* the uncertainty, but this would correspondingly reduce the practical utility of the measurement result. In chapter 4 these matters will be considered more carefully.

2.5 Measurement error, accuracy and precision

Before proceeding to study ways to evaluate the measurement uncertainty, it is very important to distinguish clearly between the concepts of uncertainty and error. In the VIM, *error* is defined as a result of a measurement minus a true value of a measurand. In turn, *a true value* is defined as a value that is consistent with the definition of the measurand. The indefinite article 'a' rather than the definite article 'the' is used in conjunction with 'true value', since there may be many values consistent with the definition of a given particular quantity (for example, many values would be consistent with a measurand defined as 'the diameter of a 1 euro coin').

True value is equivalent to what hitherto we have been calling *actual value*. Both adjectives, true and actual, are in fact redundant. One could as well suppress them and understand that *a value* consistent with the definition of a measurand quantifies what that measurand 'measures'. The problem is, of course, that such a value cannot be known. It can only be *estimated* by measurement, the result having always an associated non-zero uncertainty, however small.

In principle, a (true or actual) value could only be obtained by a 'perfect' measurement. However, for all practical purposes, a traceable best estimate with a negligible small uncertainty can be taken to be equal to the value of the measurand (especially when referring to a quantity realized by a measurement standard, such an estimate is sometimes called *conventional true value*). To illustrate this, assume that a rod's length is measured not with a caliper, but with a very precise and calibrated (traceable) micrometre. After taking into account all due corrections the result is found to be 9.949 mm with an uncertainty smaller than 1 μm. Hence, we let the length 9.949 mm be equal to the conventional true value of the rod.

Now comes the inspector with his/her caliper and obtains the result 9.91 mm. The error in this result is then equal to $(9.91 - 9.949)$ mm $= -0.039$ mm. A second measurement gives the result 9.93 mm, with an error of -0.019 mm. After taking a large number of measurements (say 10) the inspector calculates their average and obtains 9.921 mm, with an error of -0.028 mm (it is reasonable to write 9.921 mm instead of 9.92 mm, since the added significant figure is the result of an averaging process). The value -0.028 mm is called the *systematic error;* it is defined as the mean that would result from a very large number of measurements of the same measurand carried out under repeatability conditions minus a true value of a measurand (figure 2.12).

Note that the systematic error is very close to the *negative* of the correction, which by calibration was determined as being $+0.029$ mm. This is how it should be, because an error is *expected* to be eliminated by adding the corresponding correction to the uncorrected result. Systematic errors are caused by what we will call *systematic effects:* influences that lead to biases. These biases need to be corrected in order to get closer, hopefully, to the (true or actual) value of the measurand.

Return now to the individual measurement results. The first was 9.91 mm. With respect to the average, its error is $(9.91 - 9.921)$ mm $= -0.011$ mm. This value is called a *random error;* it is defined as a result of a measurement minus the mean that would result from a very large number of measurements of the same

Measurement error, accuracy and precision   37

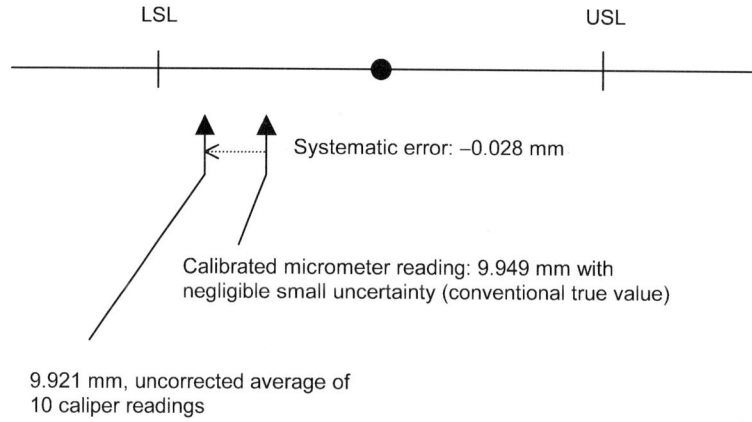

Figure 2.12. Systematic error.

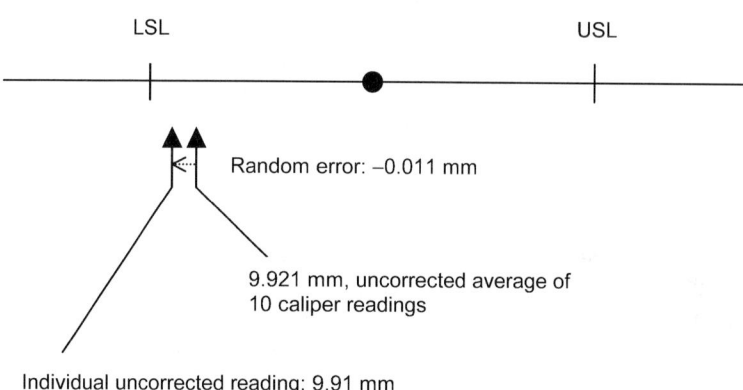

Figure 2.13. Random error.

measurand carried out under repeatability conditions (figure 2.13). Therefore, the error for an *individual* measurement result is equal to the sum of the random error and the systematic error, in this case, −0.039 mm (figure 2.14).

Admittedly, this is not only confusing but quite useless. The evaluation of the systematic, random and total errors depended on our knowledge of the (conventional) true value of the rod's length. But, of course, the inspector does not know this value. If he/she did, he/she would not have had to measure the rod with a caliper in the first place!

In summary, errors cannot be known. Systematic errors (or rather, their negatives) can only be *estimated* through the corresponding correction, but always with an attached uncertainty. Thus, uncertainty and error are very different concepts: the former is a range of values, while the latter is a point estimate.

38   *Basic concepts and definitions*

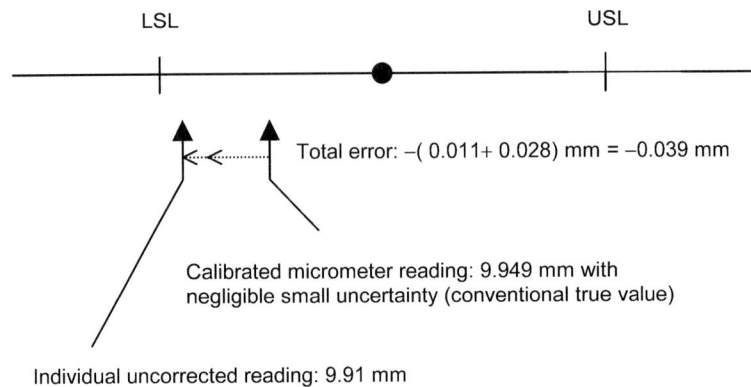

Figure 2.14. Total error.

Unknowingly, a measurement result can have a very small error and a large associated uncertainty. The reverse is also true: a measurement result may carry with it a very small uncertainty while its (unknown) error is large. However, this would mean that at least one important systematic effect must have been overlooked in the evaluation process.

Error and uncertainty are quite often taken as synonyms in the technical literature. In fact, it is not uncommon to read about 'random' and 'systematic' uncertainties. The use of such terminology is nowadays strongly discouraged.

It is also important to define the related concepts of *accuracy* and *precision*. The former is the closeness of the agreement between the result of a measurement and a true value of the measurand. Thus, a measurement is 'accurate' if its error is 'small'. Similarly, an 'accurate instrument' is one whose indications need not be corrected or, more properly, if all applicable corrections are found to be close to zero. Quantitatively, an accuracy of, say, 0.1 %—one part in a thousand—means that the measured value is not expected to deviate from the true value of the measurand by more than the stated amount.

Strangely enough, 'precision' is not defined in the VIM. This is probably because many meanings can be attributed to this word, depending on the context in which it appears. For example, 'precision manufacturing' is used to indicate stringent manufacturing tolerances, while a 'precision measurement instrument' might be used to designate one that gives a relatively large number of significant figures. In section 3.8 we take the term 'precision' as a quantitative indication of the variability of a series of repeatable measurement results. The conceptual distinction between accuracy and precision is illustrated graphically in figure 2.15.

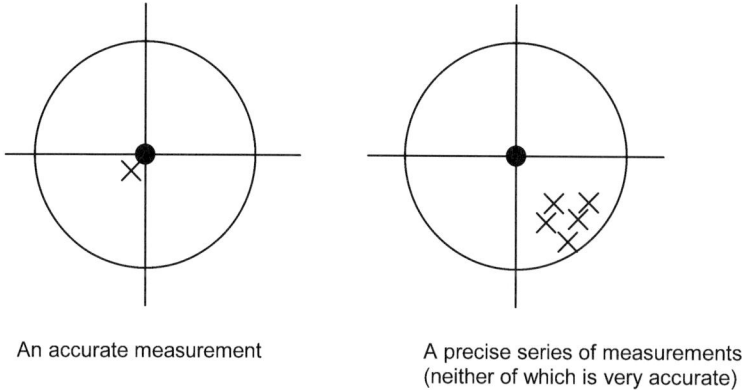

An accurate measurement   A precise series of measurements
(neither of which is very accurate)

Figure 2.15. Accuracy and precision.

2.6 The measurement model

In any measurement one must clearly and unambiguously state the measurand: the final quantity of interest. Many times the information about the measurand is acquired from the readings of a single measuring instrument. We will refer to such a quantity as subject to a *direct* measurement. However, the metrological meaning of the word measurement is more ample: it also refers to quantities whose values are *indirectly* estimated on the basis of the values of other quantities, which in turn may or may not have been directly measured.

In general, to estimate the values of measurands subject to an indirect measurement a proper *measurement model* needs to be established. The model should provide, to the best of one's knowledge, a realistic picture of the physics involved in the measurement. The model may be very simple or extremely complicated. It may be written explicitly, generally in terms of one or more mathematical formulae or may be in the form of an algorithm.

Example. The modulus of elasticity E of a sheet metal is usually determined according to the model $E = \sigma/\varepsilon$, where σ stands for stress and ε for strain. A specimen is produced and is placed in a tensile machine. The machine applies a load L that is measured directly with a transducer such as a load cell. At the same time, the distance ℓ between two marks on the stretched specimen is measured directly with an extensometer. The stress is then $\sigma = L/A$, where A is the cross section of the specimen, and the strain is $\varepsilon = (\ell - \ell_o)/\ell_o$, where ℓ_o is the original distance between the marks. In turn, the area is evaluated according to the model $A = ae$, where a is the width of the specimen and e its thickness (figure 2.16). Thus, in this example the evaluation task has been broken down into submodels, for which the

40   Basic concepts and definitions

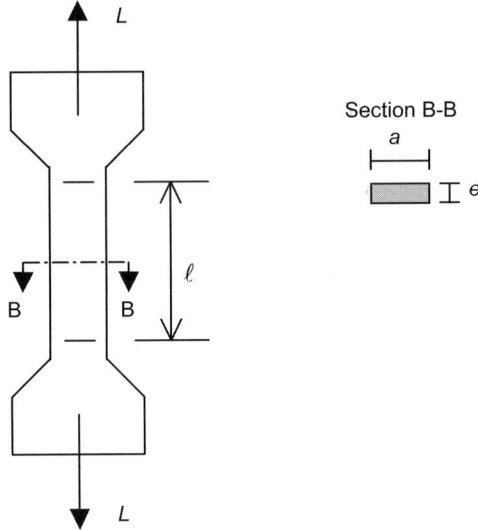

Figure 2.16. A specimen to measure the modulus of elasticity of a sheet metal.

quantities subject to direct measurement are L, ℓ, ℓ_o, a and e, while A, σ and ε are intermediate quantities (figure 2.17). Alternatively, one can combine these submodels into a single model whose input quantities are only those directly measured. This would be $E = L\ell_o/[ae(\ell - \ell_o)]$ (figure 2.18).

In general, the measurement model may be thought of as a black box that takes in *input quantities* and produces *output quantities*. The primary input quantities need not be those directly measured by the evaluator: he or she can make use of values of quantities that have been determined on a different occasion, possibly by other observers. Also, the output quantities are not necessarily the measurands of interest, because they may in turn act as input quantities to another model. For example, the modulus of elasticity E of the sheet metal might be required for calculating the deformation of a beam under an applied load, and there may be no need to measure E if its value can be estimated from data in the literature. We refer to a quantity whose value is not estimated as part of the current evaluation task as an *imported* quantity.

Even the simplest model will be incomplete if corrections to the indications of the instruments used in direct measurements are not taken into account. Indeed, every directly measured quantity Q should be modelled as $Q = G + C$, where G is the quantity associated with the gauged indications of the instrument and C is a quantity that consists of one or more correction quantities. Alternatively, one may write $Q = FG$, where F is a correction factor. Even though the estimated value of C will often be zero (or that of F will be one) for the measurement result

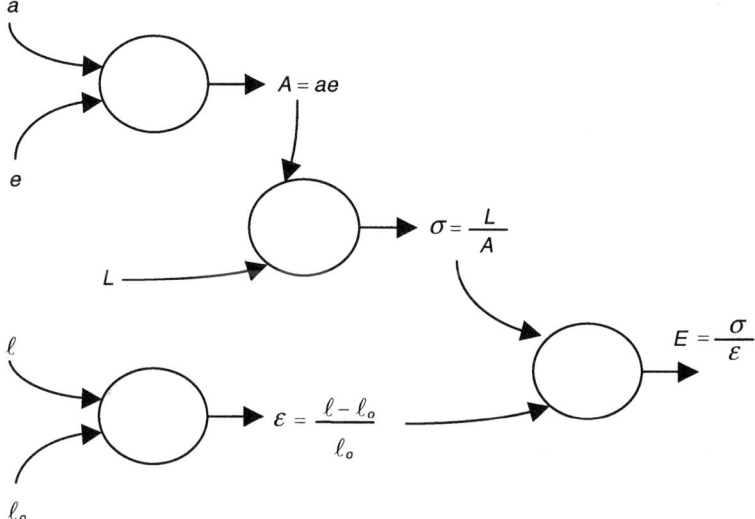

Figure 2.17. Combined submodels.

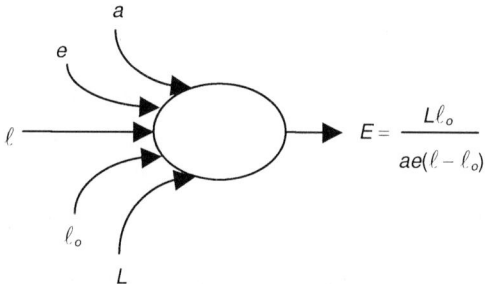

Figure 2.18. Single model.

to be traceable one should state that this information was in fact used to make the calculations. Note that, in this sense, no measurement can strictly be considered to be 'direct'.

The estimated value of the measurand is obtained by simply replacing in the model the estimates of the input quantities. In the previous example, if the stress is estimated as 170 MPa and the strain as 0.2 %, the estimated value of the modulus of elasticity becomes $170/0.002$ MPa $= 85$ GPa.

But this is just the first part of the evaluation task; the second part consists of assessing the measurement uncertainty. This is done by combining the uncertainties associated with the estimated values of the input quantities. The way to do this is prescribed by the model and is rather easy in most cases,

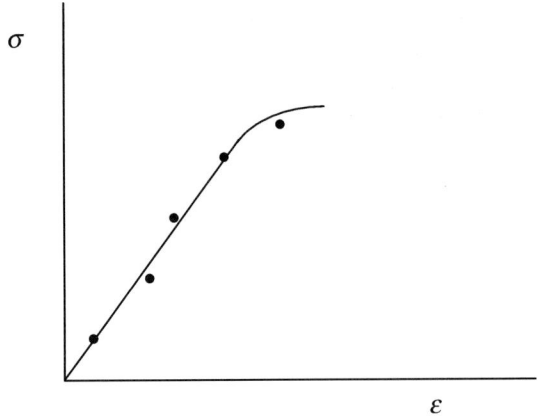

Figure 2.19. Results of a tension test plotted on a stress–strain diagram.

but there are situations that are not so simple. This happens when the model itself is complicated. Another complication arises if the input quantities are correlated. For example, if the input quantities a and e in the model $A = ae$ have been measured with a common instrument, such as a caliper, the correction to its indications induces a correlation between a and e. Even if the value of this correction is taken as zero, its uncertainty will have an influence on the uncertainty associated with the estimate of A. A further complication is due to the fact that some quantities may not appear explicitly in the model, even though they do modify the uncertainty. Finally, the simplified procedure to evaluate the uncertainty, to be expounded in section 3.4, assumes that the linear expansion of the model around the estimated values of the input quantities is accurate. However, there are cases where higher-order terms cannot be neglected for the measurement result to be meaningful. An example will be given in section 3.5.

However, the evaluator must obviously make sure that the model is correct and that it applies to the estimated values of the input quantities. Otherwise, the model will still produce values for the measurand and for its uncertainty, but these results will be meaningless. For example, the model $A = ae$ does not apply to a specimen whose cross section is known not to be rectangular. Similarly, it should be verified that the deformation of the specimen at the instant corresponding to the measurement of the load L is within the elastic range of the material. In fact, most tension tests proceed by taking a series of data points and plotting the results on a stress–strain diagram (figure 2.19). The modulus of elasticity is then obtained as the slope of the initially straight segment of the plot. For this, however, a more refined measurement model is needed.

Now, of course, the model itself may be uncertain or may be plainly but unknowingly wrong. It is here that the benefits of performing *intercomparison measurements* are most apparent (sections 1.6.4, 5.8.1 and 6.5.4). If the same measurand is measured (under reproducibility conditions) by different laboratories, and if the results are significantly different, they may indicate that at least one of those laboratories has used a model that failed to include one or more quantities that represent unrecognized systematic effects. In that case, possibly, the stated uncertainties should be enlarged, but only after all relevant data and models have been critically inspected. In other words, the procedure of arbitrarily adopting a safety factor to produce a 'safer' uncertainty should be used with extreme care, if ever.

2.7 Some additional remarks

In this chapter, the uncertainty of measurement was discussed in relation to its importance for conformity assessment. However, it arises everywhere a measurement is performed. In point of fact, evaluating the uncertainty is equivalent to quantifying the lack of complete knowledge about a particular measurand. The process should put in correspondence larger uncertainties with lesser amounts of information. Since the assessment is often conducted on the basis of estimated values, e.g. maximum error bounds or standard deviations of experimentally observed frequency distributions, it is usual to refer to an uncertainty as having been 'estimated'. Such an expression will be avoided in this book, because it communicates the mistaken idea that there is a 'true' uncertainty and that it acts as a sort of measurand whose hidden value is there for us to find.

We will also refrain from referring to the 'uncertainty *of* a measurement result'. There are two reasons for this. First, a measurement result should be understood as comprising *both* the estimated value of the measurand and its associated uncertainty. Second, once the estimated value is obtained, there is nothing uncertain about it. For example, from the information that $\sigma = 170$ MPa and $\varepsilon = 0.2\%$, we are sure that the value $E = 85$ GPa follows from the model $E = \sigma/\varepsilon$. What is uncertain is whether this value is effectively equal to the modulus of elasticity of the material. Hence, expressions such as the uncertainty *in knowing the measurand* or the uncertainty *associated with an estimated value* are more appropriate, even if longer sentences result.

The value of the measurand is usually assumed to be invariant while it is being measured. Of course, this is not always the case. If it is known that a transient or oscillating phenomenon is being measured, this fact should be duly incorporated into the model, especially if repeated measurements are made. Another assumption, usually implicit, is that the unknown value of the measurand is unique. This is sometimes emphasized through the somehow redundant concept of 'true' value. As we saw in section 2.5, this notion is used to distinguish the value of the measurand from its estimated value, which may not be 'true' not only because of the measurement uncertainty but also because some systematic effects may not have been corrected. We feel, however, that in most cases speaking about 'the' value of the measurand and adding the word *estimated* to what is obtained from measurement is enough to convey the difference between the two. Therefore, in the few occasions later where we will refer to a true value, it will be done only for emphasis.

Finally, a comment on quality. In the introductory paragraph to this chapter, it was stated that the uncertainty can be regarded as a quantitative indication of the quality of a measurement result. At first sight this is intuitively correct: if two results of the same quantity are available, the one having a smaller uncertainty will be better than the other. However, by itself the uncertainty says nothing about the *care* put into modelling the measurand, performing the actual measurements and processing the information thus obtained. For example, a small uncertainty may be due to the fact that some important systematic effect was overlooked. Hence, the quality of a measurement can be judged on the basis of its stated uncertainty solely if one is sure that every effort has been taken to evaluate it correctly.

Chapter 3

Evaluating the standard uncertainty

The uncertainty of measurement can be assessed and expressed in many ways. In view of the lack of international consensus that used to exist on this matter, the world's highest authority in metrology, the *Comité International des Poids et Mesures* (CIPM), decided that a general and accepted method of evaluation was needed. A Working Group on the Statement of Uncertainties, composed from delegates of various National Metrology Institutes, was then convened by the *Bureau International des Poids et Mesures* (BIPM) to propose specific advice on this subject. The result was the following 'Recommendation INC-1. Expression of Experimental Uncertainties':

1. The uncertainty in the result of a measurement generally consists of several components which may be grouped into two categories according to the way in which their numerical value is estimated:
A. those which are evaluated by statistical methods,
B. those which are evaluated by other means.

There is not always a simple correspondence between the classification into categories A or B and the previously used classification into 'random' and 'systematic' uncertainties. The term 'systematic uncertainty' can be misleading and should be avoided.

Any detailed report of uncertainty should consist of a complete list of the components, specifying for each the method used to obtain its numerical value.

2. The components in category A are characterized by the estimated variances s_i^2 (or the estimated 'standard deviations' s_i) and the number of degrees of freedom v_i. Where appropriate the covariances should be given.

3. The components in category B should be characterized by terms u_j^2, which may be considered approximations to the corresponding variances, the existence of which is assumed. The terms u_j^2 may be treated like variances and the terms u_j like standard deviations. Where appropriate, the covariances should be treated in a similar way.

45

4. The combined uncertainty should be characterized by the numerical value obtained by applying the usual method for the combination of variances. The combined uncertainty and its components should be expressed in the form of 'standard deviations'.

5. If for particular applications it is necessary to multiply the combined uncertainty by a factor to obtain an overall uncertainty, the multiplying factor must always be stated.

This Recommendation was a brief outline rather than a detailed prescription. Consequently, the CIPM asked the International Organization for Standardization (ISO) to develop a detailed guide, because ISO could more properly reflect the requirements stemming from the broad interests of industry and commerce. The ISO Technical Advisory Group on Metrology (TAG 4) was given this responsibility. It, in turn, established Working Group 3 and assigned it the following terms of reference:

> to develop a guidance document based upon the recommendation of the BIPM Working Group which provides rules on the expression of measurement uncertainty for use within standardization, calibration, laboratory accreditation, and metrology services.

The outcome was the ISO *Guide to the Expression of Uncertainty in Measurement* [21], hereafter the *Guide*, on which the present and the next chapter are based. However, before proceeding, it should be mentioned that, at the time of this writing, a Joint Committee for Guides in Metrology (JCGM) is now active in support of the *Guide* and the VIM via guidance documents and revisions. Working Group 1 (WG1) of the JCGM, entitled Measurement Uncertainty, is concerned with the maintenance and revision of the *Guide*. It is currently preparing two supplemental guides, one on the propagation of distributions and the other on the classification of measurement models and their evaluation using matrix analysis (this topic is the subject of chapters 5 and 6). It is expected that WG1 will also issue guidance documents on conformance testing and on measuring modelling, as appropriate to users in industrial metrology. Working Group 2 (WG2) of the JCGM is concerned with a revision of the VIM.

Both the *Guide* and the VIM define formally 'uncertainty of measurement' as

> *a parameter, associated with a result of a measurement, that characterizes the dispersion of the values that could reasonably be attributed to the measurand.*

To transform this general definition into an *operational* way of uncertainty evaluation, measurable quantities are looked from the point of view of the theory of probability. Therefore, a brief review of the concepts in that theory is in order (section 3.1). The definitions for standard, mutual and expanded uncertainties is given next (section 3.2), to be followed by a brief introduction to some common probability density functions (section 3.3). The important 'law of propagation of uncertainties' is then derived in section 3.4 for the case of only one output quantity. This law applies only to linear models. For nonlinear models one may apply the procedures described in sections 3.5 or 3.6. We turn next to the evaluation of primary quantities, i.e. those that are imported from other sources (section 3.7) and those that are measured directly (sections 3.8 and 3.9). Finally, correction quantities are defined and their way of evaluation is described in sections 3.10 (for systematic effects) and 3.11 (for calibration corrections).

3.1 Quantities as random variables

Classically, probabilities are linked to *events*, defined as the possible outcomes of a *random experiment*. This is a situation that—at least in principle—is capable of being repeated indefinitely under essentially unchanged conditions without the observer being able to predict with certainty what the outcome will be. Consider now the set of all possible events of a random experiment. This is called the *sample space*. A function that assigns a real number to each element of the sample space is known as a *random variable* (note the unfortunate use of words: a random variable is in fact a function!). The range of this function, that is, the set of all possible values of the random variable, is known as the *range space*. We now define a *probability function* whose domain is the range space of the random variable and whose range is the interval (0, 1), such that 1 is assigned to values of the random variable known to occur and 0 to values that do not occur. Figure 3.1 illustrates these concepts.

> **Example.** Let a random experiment be the toss of two dice. The sample space consists of the 36 events (n_1, n_2), where n_1 and n_2 take the values 1 through 6. Several random variables may now be defined, for example, $X = n_1$, $Y = n_1 + n_2$ or $Z = n_1 n_2$. Evidently, for each of these variables, the probability of obtaining negative values is 0, and that of obtaining values less than 37 is 1.

In the applications on which we are interested, the 'experiment' is the measurement of some quantity Q. This is taken as a random variable equal to the identity function. In this case there is no distinction between sample space and range space. Both spaces are continuous in principle and consist of all possible outcomes of the measurement. These outcomes will be denoted as q. We define a *probability density function*, referred to in what follows as pdf, as the function $f(q)$ into the range $(0, \infty)$ such that the infinitesimal probability dp that the value of the random variable falls between the values q and $q + dq$ is $f(q)\, dq$. It follows that the probability p of the random variable taking on values between two given limits q_a and q_b (figure 3.2) is equal to

$$p = \int_{q_a}^{q_b} f(q)\, dq. \tag{3.1}$$

48   Evaluating the standard uncertainty

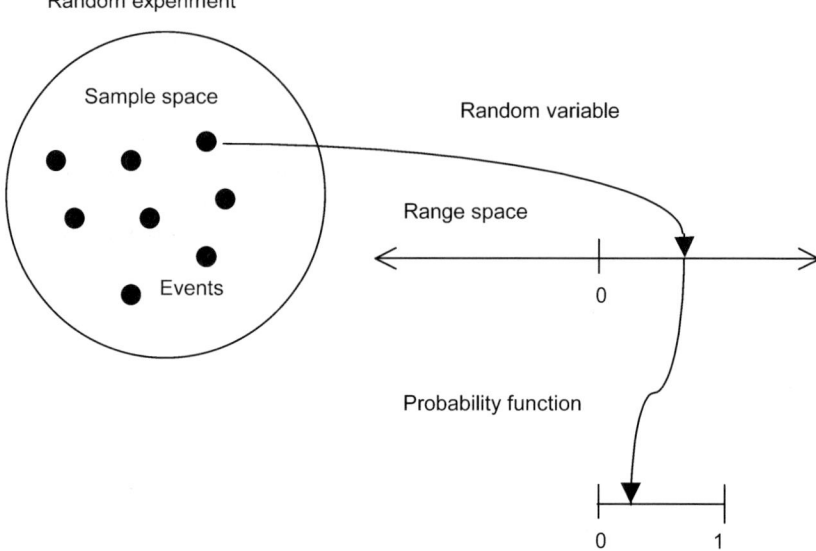

Figure 3.1. Schematic representation of a random variable.

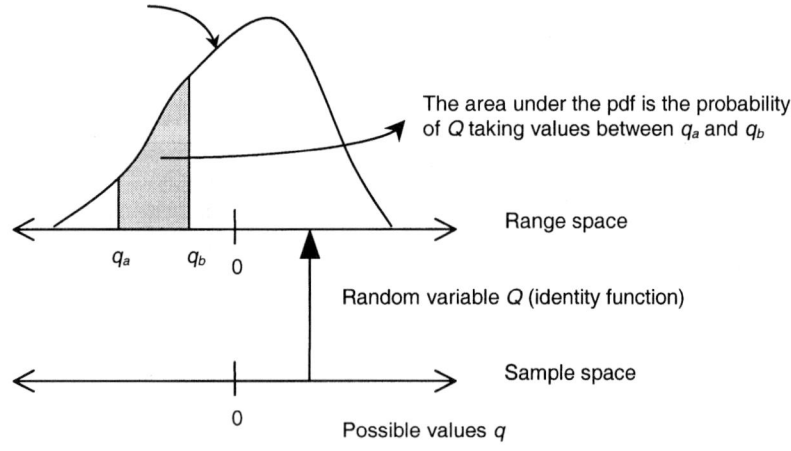

Figure 3.2. Probability density function associated with the quantity Q, interpreted as a random variable.

Since one of the values in range space must occur as a result of the measurement, it follows that, irrespective of the form of the pdf, it must be

$$1 = \int_{-\infty}^{\infty} f(q)\,dq. \tag{3.2}$$

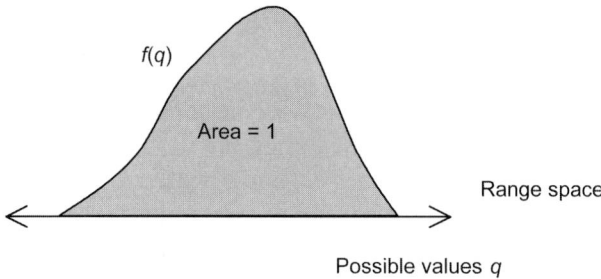

Figure 3.3. Normalization condition.

This is known as the *normalization* condition (figure 3.3). Note that, if the pdf turns out to be zero everywhere except over some interval, the limits of the integral in (3.2) may be replaced for convenience by those of the interval.

Probabilities are identified classically with the relative frequency with which the elements of the sample space would occur after an infinite number of repetitions of the random experiment. For example, in the earlier die-tossing experiment, the reader can verify that $P(X = 12) = 0$, $P(Y = 12) = \frac{1}{36}$ and $P(Z = 12) = \frac{1}{9}$, where P stands for 'probability of'. In this book, however, we will use a broader interpretation of probability. In agreement with common sense, it will be taken as a quantitative description of the likeliness, plausibility or degree of belief about a given proposition, for example, that the value of the quantity is within a prescribed interval (see the discussion in Clause E.3.5 of the *Guide*). The degree of belief depends upon the *state of knowledge* about the proposition, which is constructed in turn on the basis of all available information about the quantity, either acquired from measurement or otherwise. Thus, if the information comes from a large number of repeated measurements the pdf will be identified with the frequency distribution of the outcomes. But any other source of information might as well be used, including personal subjective belief. Therefore, a probability *density* function is not necessarily tied to an empirical *frequency* distribution of values. In this book, both concepts will always be clearly distinguished.

3.2 Standard, mutual and expanded uncertainties

The pdf associated with any measurable quantity describes not reality itself, but our knowledge about reality. The pdf should agree with and convey all the information that pertains to the quantity. We shall delay until later a discussion of the ways in which such a construct can be produced. Now suppose the pdf is finally available. It is obviously not practical to state it as the outcome of the measurement, because most users would not know what to do with it. For them, two of the pdf's most important parameters would often suffice: the *expectation* (also called the expected value) and the *variance*. They are defined in section 3.2.1.

If our interest lies in two quantities knowledge of which cannot be stated independently, the appropriate construct is called the *joint* pdf, for which the important parameter is the *covariance*. This is defined in section 3.2.2.

In many cases, however, the user will not be much interested in mathematical parameters. All he or she needs is a range of values that very probably contains the true value of the measurand. A complete discussion of this point will be presented in chapter 4, but an introduction to it is given in section 3.2.3.

3.2.1 Best estimate and standard uncertainty

Given a quantity Q and its attached pdf $f(q)$, the expectation $E(Q)$ and the variance $V(Q)$ are defined as

$$E(Q) = \int_{-\infty}^{\infty} q f(q) \, dq$$

and

$$\begin{aligned} V(Q) &= E[Q - E(Q)]^2 \\ &= \int_{-\infty}^{\infty} [q - E(Q)]^2 f(q) \, dq \\ &= E(Q)^2 - E^2(Q) \end{aligned} \quad (3.3)$$

where

$$E(Q)^2 = \int_{-\infty}^{\infty} q^2 f(q) \, dq. \quad (3.4)$$

(One might as well write $E(Q^2)$ instead of $E(Q)^2$. However, the former notation gets complicated when Q is replaced by some other expression as in equation (3.7). The reader is warned not to confuse $E(Q)^2$ with $E^2(Q) = [E(Q)]^2$.)

Among the properties of the expectation and variance operators, the following are important:

$$E(a + bQ) = a + bE(Q) \quad (3.5)$$

and

$$V(a + bQ) = b^2 V(Q) \quad (3.6)$$

where a and b are constants. Thus, the expectation is a linear operator, while the variance is not.

Following the ideas of Gauss and Legendre [41], the best estimate of Q is determined by minimizing

$$E(Q - q')^2 = V(Q) + [E(Q) - q']^2 \quad (3.7)$$

Standard, mutual and expanded uncertainties

with respect to q'. Obviously, the minimum occurs for $q' = \mathrm{E}(Q)$. We will denote this best estimate as q_e. Thus,

$$q_e = \mathrm{E}(Q). \tag{3.8}$$

Recall now the definition of the measurement uncertainty on page 46. The 'dispersion of the values' to which it alludes is fully characterized by the variance of the pdf. However, although the variance is of fundamental importance, its dimensions are those of Q squared. This fact makes it inconvenient to state the variance as the measurement uncertainty. Rather, the uncertainty is taken as the positive square root of the variance. This parameter is called the *standard deviation*, and will be denoted as $\mathrm{S}(Q)$. We then define the *standard uncertainty* as

$$u_q = \mathrm{S}(Q)$$

or, equivalently,

$$u_q^2 = \mathrm{V}(Q). \tag{3.9}$$

(Occasionally, the symbol $u(Q)$ will be used instead of u_q.)

The numerical value

$$u_{rq} = \frac{u_q}{|q_e|}$$

is called the *relative* standard uncertainty. It has no units, and the absolute sign in the denominator is needed to preclude the possibility of u_{rq} becoming negative (by definition, the standard uncertainty is always positive). Of course, the definition of the relative standard uncertainty applies only if the estimated value does not vanish.

Substitution of (3.8) and (3.9) in (3.3) gives

$$q_e^2 + u_q^2 = \mathrm{E}(Q)^2. \tag{3.10}$$

For a given pdf, the right-hand side in this equation is equal to a constant defined by (3.4). Therefore, the best estimate and the standard uncertainty form a related pair of values. However, it is sometimes more convenient to estimate the value of the quantity by a value q^* that is not equal to q_e. To characterize the dispersion of the values that could reasonably be attributed to Q *relative to* q^* we use not $\mathrm{V}(Q)$ but $\mathrm{E}(Q - q^*)^2$, and define the latter expression as the square of the standard uncertainty u_q^* associated with q^*. Then, from (3.7)

$$u_q^{*2} = u_q^2 + (q_e - q^*)^2. \tag{3.11}$$

This equation—not given explicitly in the *Guide*—shows that u_q^* is always enlarged with respect to u_q and that, since

$$(q_e - q^*)^2 \le u_q^{*2}$$

the best estimate q_e will always be contained within the interval defined by the limits $q^* \pm u_q^*$ (figure 3.4). For an application of (3.11) see [28].

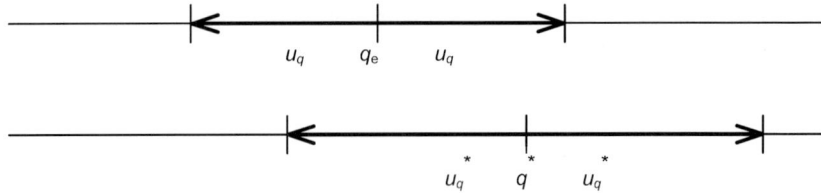

Figure 3.4. Best estimate q_e, another estimate q^* and associated standard uncertainties u_q and u_q^*.

3.2.2 Mutual uncertainties

Consider now two quantities Q and Z. In general, the information pertaining to these quantities is described by a *joint* probability density function, whose expression is

$$f(q, z) = f(z|q) f(q) \tag{3.12}$$

where

$$f(q) = \int_{-\infty}^{\infty} f(q, z) \, dz$$

is the *marginal* density for Q and $f(z|q)$ is the *conditional* density for Z given that the value of Q is q. Equivalently, the joint density can be expressed as the product of the marginal density for Z multiplied by the conditional density for Q given that the value of Z is z:

$$f(q, z) = f(z) f(q|z).$$

In these expressions, the same symbol (f) represents *different* densities. Strictly, one should be more specific and write for example $f(q)$, $g(q, z)$ and so forth. This will not normally be necessary, as the argument or arguments of the various pdfs will make it clear to which quantities they refer and whether they are marginal, joint or conditional densities.

The best estimates and standard uncertainties of Q and Z can be obtained as the expectations and standard deviations of their respective marginal pdfs. An alternative procedure is to use the formula (see e.g. [45, p 97])

$$V(Q) = E_Z[V(Q|Z)] + V_Z[E(Q|Z)] \tag{3.13}$$

where the subscript on the right-hand side indicates that the expectation value and the variance are taken with respect to the quantity Z. A similar equation holds for the variance of Z. For an example on the use of (3.13) see section 3.3.3.

Let us now define the *covariance* between the quantities Q and Z as

$$\begin{aligned} C(Q, Z) &= E\{[Q - E(Q)][Z - E(Z)]\} \\ &= \int_{-\infty}^{\infty} \int_{-\infty}^{\infty} [q - E(Q)][z - E(Z)] f(q, z) \, dq \, dz. \end{aligned} \tag{3.14}$$

This parameter is taken as being equal to the *mutual uncertainty* between the quantities. Our notation will be

$$u_{q,z} = C(Q, Z).$$

The dimensions of the mutual uncertainty are equal to the product of the dimensions of each quantity. As this may be confusing, the *correlation coefficient* is a more convenient parameter. This is defined as

$$r_{q,z} = \frac{u_{q,z}}{u_q u_z} \tag{3.15}$$

it is dimensionless and its value is always in the interval $(-1, 1)$.

The quantities Q and Z are said to be *correlated* if their covariance does not vanish. They are *independent* if their joint pdf can be written as the product of their respective marginals

$$f(q, z) = f(q)f(z).$$

From (3.14) it may be seen that independent quantities are also uncorrelated. However, the converse is not necessarily true, that is, it is possible that $C(Q, Z) = 0$ while the quantities are not independent.

3.2.3 The expanded uncertainty

In practice, the users of a measurement result will often not be interested in the standard uncertainty *per se*. For decision purposes, for example in conformity assessment, they require an interval I_p that can be viewed to contain the value of the measurand with a relatively high *coverage* probability p. This probability is referred to in the *Guide* as the 'level of confidence'†. By writing $I_p = (q_a, q_b)$, equation (3.1) follows.

It will be shown in section 4.1 that, for symmetric pdfs, centring the interval I_p at q_e and selecting its half-width as u_q results in a relatively low coverage probability. If a higher value is required, the width of the interval has to be increased. To do so, the *Guide* recommends multiplying the standard uncertainty by an appropriate p-dependent number $k_p > 1$, known as the *coverage factor*. Rules to calculate k_p are given in the *Guide*; they will be discussed in chapter 4. Note that k_p is not an arbitrary 'safety factor', but a parameter associated with a coverage probability. Despite this fact, it is usually written without a subscript.

The product $U_{pq} = k_p u_q$ is known as the *expanded uncertainty*; it is equal to the half-width of the required (symmetric) $100p\%$ coverage interval. The complete measurement result is then commonly stated as $q_e \pm U_{pq}$, where it is assumed that the coverage probability is sufficiently high. (The *Guide*'s notation is U in its chapter 7 and U_p in its appendix G).

† 'Level of confidence' should not be confused with 'confidence level'. The latter is a concept related to sampling theory, as will be explained in section 4.5.

3.3 Some useful pdfs

Two questions now arise:

(i) Should we *always* use a pdf to evaluate a standard uncertainty?
(ii) Are there objective criteria with which to construct a 'proper' pdf $f(q)$, in the sense that the information available about the quantity Q is correctly represented by its pdf?

The answer the *Guide* gives to (i) is 'no'. Therein, procedures are recommended to obtain a standard uncertainty that is not based on the direct evaluation of expectations and standard deviations. One of these consists of applying the so-called 'law of propagation of uncertainties' to quantities that are modelled in terms of other quantities whose best estimates, standard uncertainties and mutual uncertainties are given. This law will be derived in section 3.4 for the case of only one output quantity and will be generalized in section 5.2 for several inter-related output quantities. Another method—totally different in its basic philosophy—is based on *sampling theory* to estimate the standard deviation of a frequency distribution of values. This second method is termed in the *Guide* the 'type A evaluation of uncertainty' and applies to quantities that are measured directly. It will be studied in section 3.8.

The answer to (ii) will be deferred to chapter 6 and, in particular, to section 6.6. For the moment, however, it is convenient to introduce two pdfs that appear frequently in metrological applications and can be used on rather intuitive grounds. These are the normal (or Gaussian) and the rectangular (or uniform) pdfs. The *Guide* gives also some attention to the trapezoidal pdf— again on an intuitive basis—and for this reason it is presented here. However, the discussion in section 6.6 will lead to the conclusion that the trapezoidal pdf is not as fundamental as the normal and rectangular pdfs. In some cases, a better alternative to the trapezoidal pdf might be the logarithmic pdf. This is derived in section 3.3.3.

3.3.1 The normal pdf

The normal pdf is

$$N(q) = \frac{1}{\sqrt{2\pi}\sigma} \exp\left[-\frac{1}{2}\frac{(q-\mu)^2}{\sigma^2}\right] \qquad (3.16)$$

for possible values q that extend over the real line, i.e. from $-\infty$ to ∞ (figure 3.5).

In this expression, the parameter μ corresponds to the expectation value and the parameter σ^2 corresponds to the variance. Therefore, without the need to perform any integration, we write immediately

$$q_e = \mu$$

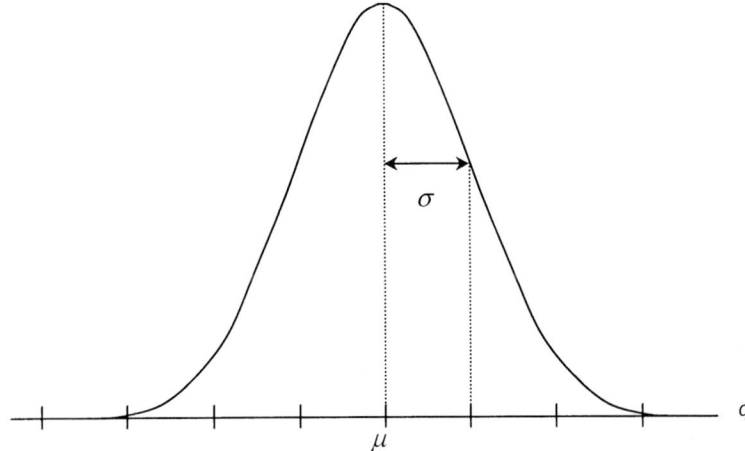

Figure 3.5. The normal (or Gaussian) pdf.

and
$$u_q = \sigma.$$

Definite integrals of the normal pdf cannot be expressed in terms of analytic functions. However, they can be evaluated numerically from the fact that if the pdf of Q is normal, then the pdf of $X = (Q - q_e)/u_q$ is also normal, but centred at $x = 0$ and unit variance:

$$N_s(x) = \frac{1}{\sqrt{2\pi}} \exp\left(-\frac{1}{2}x^2\right).$$

This is called the *standard normal* pdf. The numerical values of its *cumulative function*

$$\Phi(x) = \int_{-\infty}^{x} N_s(z)\,dz$$

have been calculated for various values x and are given in table A1 in the appendix. Therefore,

$$\int_{q_a}^{q_b} N(q)\,dq = \Phi(x_b) - \Phi(x_a) \tag{3.17}$$

where $x_a = (q_a - q_e)/u_q$ and $x_b = (q_b - q_e)/u_q$.

Example. The breaking strength Q of a sample of cotton fabric is estimated as $q_e = 165$ N with $u_q = 3$ N. Assume that these parameters apply to a normal density. To obtain the probability that the actual breaking strength is less than, say $q_b = 162$ N, we define

$X = (Q - 165)/3$, calculate $x_b = (162 - 165)/3 = -1$ and let $x_a = -\infty$ (note that the values of X are adimensional). From table A1 we find the required probability as $\Phi(x_b) = 0.1587$.

Example. The best estimate of the length Q of a certain piece of equipment is $q_e = 10.11$ mm with a coverage interval $I_{0.5} = (10.07, 10.15)$ mm. In other words, it is assumed that the probability of the length being inside this interval is equal to that of its being outside. To evaluate the standard uncertainty we assume a normal pdf centred at q_e and use the conditions $\Phi(x_b) - \Phi(x_a) = 0.5$ together with $\Phi(x_a) = 0.25$ and $\Phi(x_b) = 0.75$. The value $x_b = 0.675$ is obtained from table A1 and the standard uncertainty becomes $u_q = (q_b - q_e)/x_b = (10.15 - 10.11)/0.675$ mm $= 0.059$ mm.

The normal pdf has a very important status in statistical theory. In fact, its name is due precisely to the fact that it approaches the frequency distribution of values 'normally' occurring under some common circumstances. For example, suppose that the sample space of a random experiment consists of only two events, say A and B, which occur with probabilities p and $1 - p$ respectively. Consider n repetitions of the experiment and define the random variable Z as equal to the number of times that A occurred. The probability of Z being equal to a given integer $k \leq n$ is given by

$$\frac{n!}{k!(n-k)!} p^k (1-p)^{n-k}.$$

This is called the binomial distribution. Now define

$$Q = (Z - np)[np(1-p)]^{-1/2}.$$

It can be shown that, for large n, $f(q)$ is approximately given by (3.16).

Of more importance to metrological applications is the fact that if a quantity is measured repeatedly, its influence quantities that vary stochastically from reading to reading will produce individual (random) errors whose frequency distribution can usually be modelled according to the normal pdf with vanishing expectation. The type A evaluation of expanded uncertainty in section 4.5 takes advantage of this occurrence.

The normal pdf will never apply strictly to a quantity whose possible values are known to be restricted to some given interval (for example, from physical considerations the measurand might be known to be positive). Nevertheless, the mathematical convenience afforded by the normal pdf will, in many cases, justify its use against a more elaborate probability density.

3.3.2 The trapezoidal, rectangular and triangular pdfs

The symmetric trapezoidal pdf is depicted in figure 3.6. It consists of an isosceles trapezoid over an interval where it is known that the (true) value of the quantity Q

Some useful pdfs   57

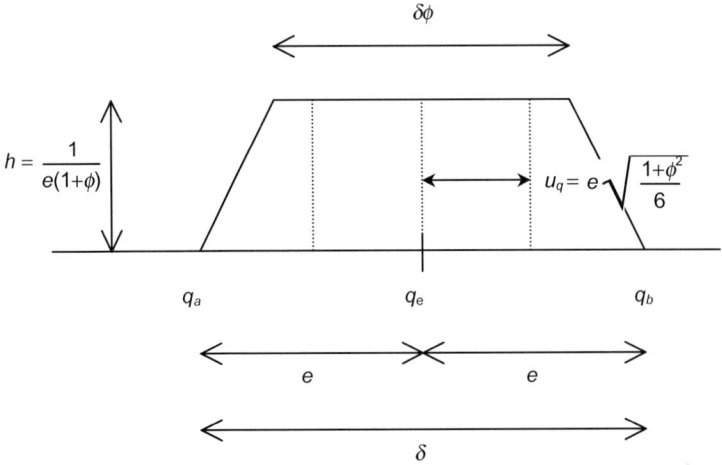

Figure 3.6. The symmetric trapezoidal pdf.

lies. Typically, this interval will be given in the form (q_a, q_b) or $q_e \pm e$, where q_e is the best estimate, e is a known maximum error, $q_a = q_e - e$ and $q_b = q_e + e$. Denote the width of the base of the trapezoid as $\delta = 2e = q_b - q_a$ and the width of the top as $\delta\phi$, where ϕ is a parameter that can take any value between zero and one. The normalization condition imposes, for the pdf, a height $h = [e(1+\phi)]^{-1}$.

Due to symmetry, there is no need to integrate the trapezoidal pdf to obtain the expectation; it is just the midpoint:

$$E(Q) = q_e = \frac{q_a + q_b}{2}.$$

To derive an expression for the variance, consider a trapezoidal pdf centred at zero with $q_a = -e$ and $q_b = e$. Since $E(Q) = 0$, from (3.3) we have

$$V(Q) = E(Q)^2 = 2\int_0^{e\phi} q^2 f_1(q)\,dq + 2\int_{e\phi}^{e} q^2 f_2(q)\,dq$$

where

$$f_1(q) = \frac{1}{e(1+\phi)}$$

and

$$f_2(q) = \frac{1}{e(1-\phi^2)}\left(1 - \frac{q}{e}\right).$$

Although tedious, the integration is quite straightforward. The result is

$$V(Q) = \frac{e^2(1+\phi^2)}{6}.$$

58   Evaluating the standard uncertainty

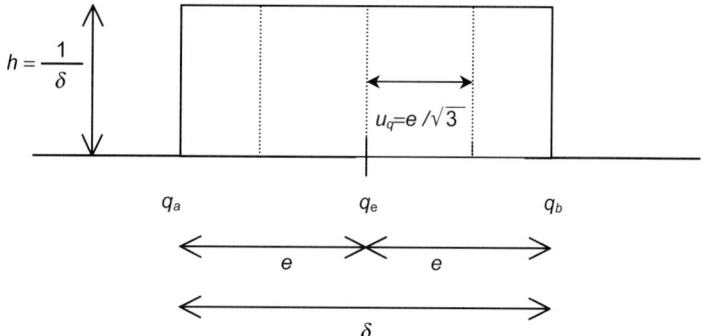

Figure 3.7. The rectangular (or uniform) pdf.

Since the variance does not depend on where the pdf is centred, this expression is general and we have

$$u_q = e\sqrt{\frac{1+\phi^2}{6}}. \qquad (3.18)$$

If $\phi = 1$ the rectangular pdf obtains (figure 3.7). Then, from (3.18),

$$u_q = \frac{e}{\sqrt{3}} = \frac{\delta}{\sqrt{12}}. \qquad (3.19)$$

If $\phi = 0$ the triangular pdf results (figure 3.8). In this case

$$u_q = \frac{e}{\sqrt{6}} = \frac{\delta}{\sqrt{24}}.$$

The rectangular pdf applies when the *only* information available about the quantity Q is that its (true) value lies within the interval (q_a, q_b). In turn, the trapezoidal and triangular pdfs apply to a quantity that is expressed as the sum of two quantities described by rectangular pdfs (see section 6.9.1).

Clause 4.3.9 of the *Guide* recommends the use of the triangular pdf in case the values near the bounds of the interval (q_a, q_b) are less likely than those near the midpoint. However, even though this recommendation appears to be intuitively acceptable, it cannot be justified rigorously. This can be seen by observing the three pdfs in figure 3.9. All of them have the same expectation and comply with the condition that values near the centre of the interval are more likely than those at the extremes. Yet, their variances are quite different. Is any one of these pdfs (or the triangular) more 'reasonable' than the others? The answer to this question will be given in section 6.6.

3.3.3 The logarithmic pdf

To obtain the logarithmic pdf we reason in the following way [29]: assume that the best estimate q_e of the quantity Q is given, together with an associated error

Some useful pdfs 59

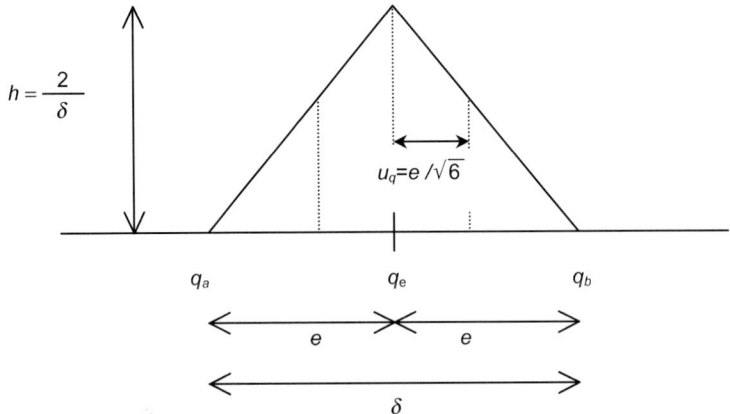

Figure 3.8. The triangular pdf.

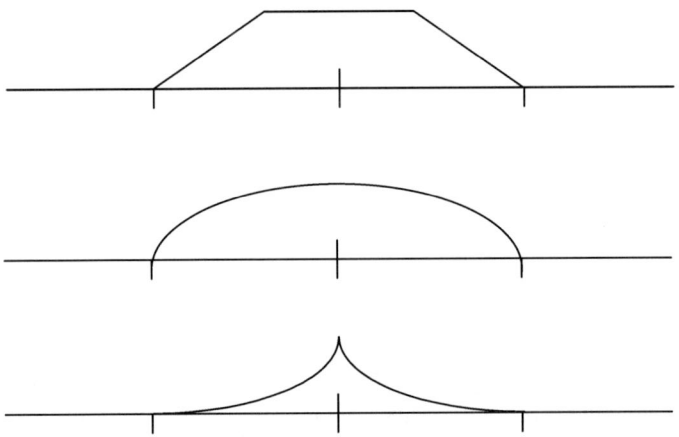

Figure 3.9. All these pdfs have the same expectation and comply with the condition that values near the centre of the interval are more likely than those at the extremes.

bound, and take this bound as an inexactly known positive quantity E. Assume also that this quantity is described by a rectangular pdf $R(e)$ centred at a given value e_e and having a known half-width ε satisfying $0 \leq \varepsilon \leq e_e$ (figure 3.10). The possible values of Q are then contained within the interval (q_a, q_b), where $q_a = q_e - e_e - \varepsilon$ and $q_b = q_e + e_e + \varepsilon$.

From (3.12), the joint probability density of Q and E is

$$f(q, e) = f(q|e) R(e)$$

where $f(q|e)$ is the conditional density for Q given that the value of E is e. Since

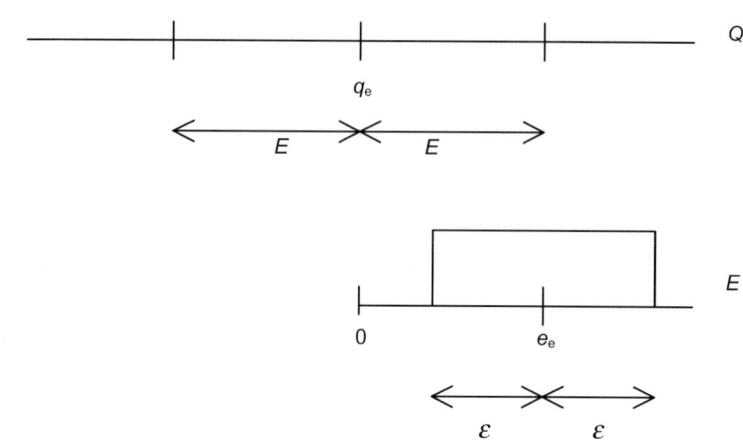

Figure 3.10. A quantity with values inside the interval $q_e \pm E$, where E is a quantity described by a rectangular pdf.

this density is also rectangular, we write

$$f(q|e) = \frac{1}{2e}.$$

Also

$$R(e) = \frac{1}{2\varepsilon}$$

and therefore

$$f(q,e) = \frac{1}{4e\varepsilon}.$$

The density $f(q)$ is now obtained by integrating $f(q,e)$ over all possible values of E. At first sight, it appears that these are contained within the interval $(e_e - \varepsilon, e_e + \varepsilon)$. However, on closer inspection one realizes that this happens only for values of Q in the interval (q_c, q_d), where $q_c = q_e - e_e + \varepsilon$ and $q_d = q_e + e_e - \varepsilon$ (figure 3.11). Instead, for $q_a < q < q_c$ and $q_d < q < q_b$, the error E cannot be smaller than $|q - q_e|$. We then obtain

$$f(q) = \begin{cases} \dfrac{1}{4\varepsilon} \ln\left(\dfrac{e_e + \varepsilon}{e_e - \varepsilon}\right) & \text{(i)} \\ \dfrac{1}{4\varepsilon} \ln\left(\dfrac{e_e + \varepsilon}{|q - q_e|}\right) & \text{(ii)} \end{cases} \qquad (3.20)$$

where (i) applies for $q_c < q < q_d$ and (ii) applies for $q_a < q < q_c$ and $q_d < q < q_b$.

The variance of the log pdf can be found by integrating (3.20). However, it is simpler (and more instructive) to use (3.13), which for this case becomes

$$V(Q) = E_E[V(Q|E)] + V_E[E(Q|E)].$$

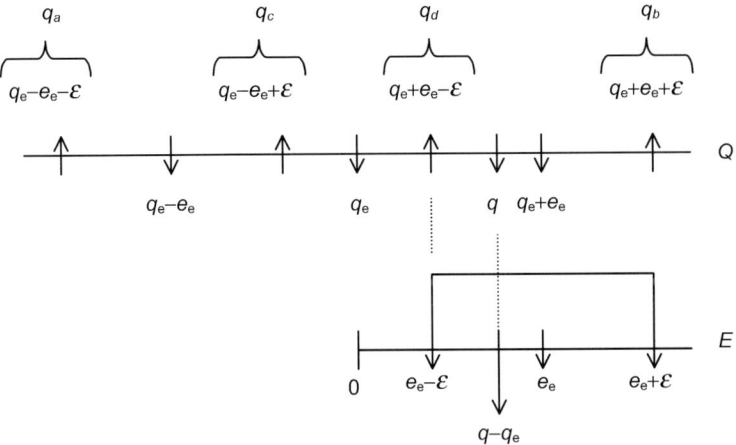

Figure 3.11. Values that define the integration intervals to obtain the log pdf.

Since for a specified value of E the conditional pdf associated with Q is rectangular, it follows that $V(Q|E) = E^2/3$ and $E(Q|E) = q_e$. Therefore, from (3.3),

$$V(Q) = E\left(\frac{E^2}{3}\right) = \frac{V(E) + E^2(E)}{3}.$$

Now $V(E) = \varepsilon^2/3$ and $E(E) = e_e$. Thus, finally,

$$u_q^2 = V(Q) = \frac{1}{3}e_e^2\left[1 + \frac{1}{3}\left(\frac{\varepsilon}{e_e}\right)^2\right]. \qquad (3.21)$$

Comparing (3.19) with (3.21), it is seen that the standard uncertainty associated with the best estimate q_e is enlarged with respect to a rectangular pdf for which the limits are defined sharply.

The log pdf is similar in shape to the trapezoidal pdf, see figure 3.12. It is also symmetrical and assigns larger probabilities to values near the best estimate than to values farther from it. However, whereas the parameter ϕ in the trapezoidal pdf has to be set rather arbitrarily, the width of the top of the logarithmic pdf is established from equation (3.20) once the two parameters e_e and ε are given.

3.4 The law of propagation of uncertainties (LPU)

Consider a quantity Q modelled explicitly in terms of a set of input quantities $Z_1, \ldots, Z_N$ according to

$$Q = \mathcal{M}(Z_1, \ldots, Z_N)$$

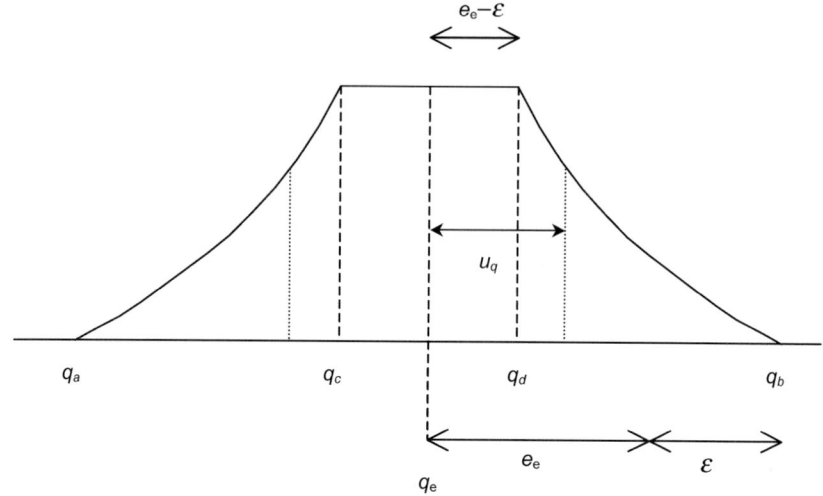

Figure 3.12. The logarithmic pdf.

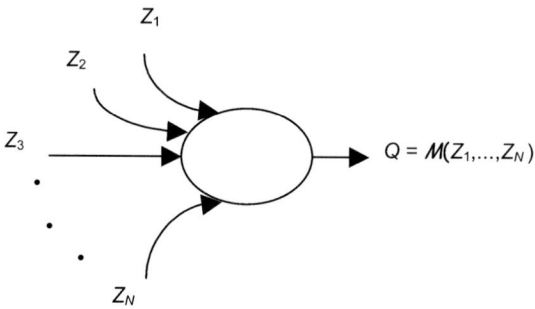

Figure 3.13. The measurement model.

where $\mathcal{M}$ stands for 'function of' (figure 3.13).

From earlier ideas, the best estimate q_e and its associated standard uncertainty u_q should be obtained as the expectation and the standard deviation of the pdf $f(q)$, respectively. This pdf can be derived from the pdfs of the input quantities. Yet, on the one hand, this will be usually difficult and, on the other hand, except in some very special cases, the integration of the resulting pdf will have to be carried out numerically.

Fortunately, there exists a rather simple procedure to obtain q_e and u_q when the estimated values $z_{ei} = \mathrm{E}(Z_i)$ are given, together with the standard uncertainties $u_{zi} = \mathrm{S}(Z_i)$. If the input quantities are correlated, their mutual uncertainties $u_{zi,zj} = \mathrm{C}(Z_i, Z_j)$ must also be known.

In effect, consider the first-order Taylor series expansion of the model around the best estimates $(z_{e1}, \ldots, z_{eN})$. This is called the *linearized* model. Its expression is

$$Q_L = \mathcal{M}(z_{e1}, \ldots, z_{eN}) + \sum_{i=1}^{N} c_i (Z_i - z_{ei})$$

$$= a + \sum_{i=1}^{N} c_i Z_i$$

where

$$a = \mathcal{M}(z_{e1}, \ldots, z_{eN}) - \sum_{i=1}^{N} c_i z_{ei}$$

and the c_is are the partial derivatives

$$c_i = \frac{\partial \mathcal{M}}{\partial Z_i}$$

evaluated at the best estimates of the input quantities. These derivatives are called the *sensitivity coefficients*.

The simplified procedure is best understood by considering the case of just two input quantities. We then have

$$Q_L = a + c_1 Z_1 + c_2 Z_2.$$

From the properties (3.5) and (3.6) and the definition (3.14), the expectation and variance of Q_L are

$$E(Q_L) = a + c_1 E(Z_1) + c_2 E(Z_2) = \mathcal{M}(z_{e1}, z_{e2})$$

and

$$V(Q_L) = c_1^2 V(Z_1) + c_2^2 V(Z_2) + 2 c_1 c_2 C(Z_1, Z_2)$$

where we have used the fact that $C(Z_1, Z_2) = C(Z_2, Z_1)$. Therefore, for the linearized quantity Q_L we have

$$q_{eL} = \mathcal{M}(z_{e1}, z_{e2})$$

and, since $u_{zi}^2 = V(Z_i)$ and $u_{zi,zj} = C(Z_i, Z_j)$,

$$u_{qL}^2 = c_1^2 u_{z1}^2 + c_2^2 u_{z2}^2 + 2 c_1 c_2 u_{z1,z2}.$$

The generalization of these expressions to N input quantities is

$$q_{eL} = \mathcal{M}(z_{e1}, \ldots, z_{eN})$$

and

$$u_{qL}^2 = \sum_{i=1}^N c_i^2 u_{zi}^2 + 2\sum_{i=1}^{N-1}\sum_{j=i+1}^N c_i c_j u_{zi,zj}.$$

Now, if the model is not a strongly nonlinear function of the input quantities, such that in the neighbourhood of the best estimates z_{ei} the quantity Q does not deviate much from Q_L, the values q_{eL} and u_{qL} can be assumed to be almost equal to q_e and u_q, respectively. Therefore, *without the need to calculate any expectation or variance*, we write

$$q_e = \mathcal{M}(z_{e1}, \ldots, z_{eN})$$

and since $u_{zi,zi} = u_{zi}^2$,

$$u_q^2 = \sum_{i,j=1}^N c_i c_j u_{zi,zj}. \qquad (3.22)$$

If all input quantities are independent, the mutual uncertainties vanish and (3.22) reduces to

$$u_q^2 = \sum_{i=1}^N c_i^2 u_{zi}^2. \qquad (3.23)$$

Both equation (3.22) and the more restricted expression (3.23) are usually referred-to as the 'law of propagation of uncertainties' (LPU) or, more properly, as the law of propagation of variances. The LPU expresses the fact that the 'uncertainty components' $c_i c_j u_{zi,zj}$ (or just $c_i^2 u_{zi}^2$ for independent input quantities) should be combined additively to obtain the square of u_q.

Example. Consider the linear, additive model

$$Q = a + \sum_{i=1}^N c_i Z_i$$

where a and the c_is are given constants. Then $Q_L = Q$ and (3.22) applies strictly, with no approximations.

Example. Consider now the multiplicative model

$$Q = a\prod_{i=1}^N (Z_i)^{b_i} = a(Z_1)^{b_1}\ldots(Z_N)^{b_N}$$

where a and the b_is are given constants. Then

$$q_e = a\prod_{i=1}^N (z_{ei})^{b_i}$$

and the sensitivity coefficients become

$$c_i = a(z_{e1})^{b_1} \ldots b_i(z_{ei})^{b_i-1} \ldots (z_{eN})^{b_N}.$$

If neither of the estimates z_{ei} vanishes, these coefficients can be written as

$$c_i = \frac{q_e b_i}{z_{ei}}.$$

Therefore, if the input quantities are independent,

$$u_q^2 = \sum_{i=1}^{N} \left(\frac{q_e b_i}{z_{ei}}\right)^2 u_{zi}^2$$

or

$$\left(\frac{u_q}{q_e}\right)^2 = \sum_{i=1}^{N} b_i^2 \left(\frac{u_{zi}}{z_{ei}}\right)^2. \qquad (3.24)$$

Thus, in models involving the products of powers of independent quantities, an expression of the same form as (3.23) applies, with the standard uncertainties replaced by the *relative* standard uncertainties and the sensitivity coefficients replaced by the exponents.

Example. Models of the form

$$Q = a \frac{Z_1 \ldots Z_n}{Z_{n+1} \ldots Z_N}$$

are very common. In this case, equation (3.24) reduces to

$$\left(\frac{u_q}{q_e}\right)^2 = \sum_{i=1}^{N} \left(\frac{u_{zi}}{z_{ei}}\right)^2. \qquad (3.25)$$

As an illustration, consider the model for the modulus of elasticity of a sheet metal. This is $E = \sigma/\varepsilon$, where σ is the stress and ε is the strain. Then, from (3.25),

$$\frac{u_E^2}{E^2} = \frac{u_\sigma^2}{\sigma^2} + \frac{u_\varepsilon^2}{\varepsilon^2}$$

or

$$u_E^2 = \frac{u_\sigma^2}{\varepsilon^2} + \frac{\sigma^2 u_\varepsilon^2}{\varepsilon^4}.$$

It should be noted that it is not always possible to write the measurement model in the form of an explicit analytic function. For example, the model may be a computer algorithm that produces the value q_e from the estimated values z_{ei}

of the input quantities. In this case the sensitivity coefficients are not obtained analytically, but numerically, by changing in turn each one of the estimated input values to $z_{ei} + \Delta z_i$, where Δz_i is as small as practical. If the corresponding value of the output quantity is subtracted from the estimate q_e, the change $-\Delta q_i$ obtains. Accordingly, $c_i \approx \Delta q_i / \Delta z_i$.

Note also that the *Guide* refers to an uncertainty calculated as per equations (3.22) or (3.23) as 'combined' standard uncertainty and gives to it a special symbol, namely $u_c(q)$. There is really no need to make this distinction.

Let us now summarize the conditions under which the LPU can be used. These are:

- Only one output quantity appears in the model.
- The model is explicit, that is, it can be written as $Q = \mathcal{M}(Z_1, \ldots, Z_N)$ (although the function $\mathcal{M}$ does not need to be analytic).
- The best estimates and the standard and mutual uncertainties of the input quantities are available.
- The model is well approximated by its linear expansion around the best estimates of the input quantities.

Chapters 5 and 6 will deal with more general measurement models involving several inter-related output quantities. The next section presents a methodology that can be applied to nonlinear models with one input quantity.

3.5 The special case of only one input quantity

The case $Q = \mathcal{M}(Z)$ can, in principle, be treated with the linearized model

$$Q_L = \mathcal{M} + \mathcal{M}'(Z - z_e)$$

where $\mathcal{M} \equiv \mathcal{M}(z_e)$,

$$\mathcal{M}' \equiv \left. \frac{d\mathcal{M}(Z)}{dZ} \right|_{Z=z_e}$$

and $z_e = E(Z)$. Then, with $u_z^2 = V(Z)$ we write

$$q_e = E(Q) \approx E(Q_L) = \mathcal{M}$$

and

$$u_q^2 = V(Q) \approx V(Q_L) = (\mathcal{M}' u_z)^2.$$

How good are these approximations? This depends on the behaviour of $\mathcal{M}(Z)$ around the best estimate z_e. For example, consider figure 3.14. The linear expansion of the depicted model around z_{e1}, Q_{L1}, coincides with $\mathcal{M}(Z)$ over a large range of values. Instead, the range of coincidence of Q_{L2} with $\mathcal{M}(Z)$ is much smaller. In fact, if the best estimate of Z is z_{e2}, $\mathcal{M}'_2$ vanishes and we get

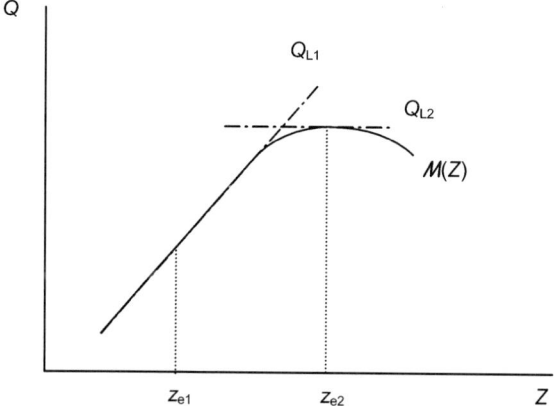

Figure 3.14. A nonlinear model.

$V(Q_{L2}) = 0$. If this result is also assumed to hold for $V(Q)$, it would imply that Q is perfectly known. In this case, the LPU fails.

Example. Consider a quantity Q that oscillates about the value $a + b$ with amplitude $b > 0$ (figure 3.15). The model is

$$Q = a + b(1 - \cos Z) \tag{3.26}$$

where Z is proportional to the time at which Q is observed. Let $z_{e1} = \pi/2$ and $u_{z1} = \pi/8$. Then $q_{e1} = \mathcal{M}_1 = a + b$, $\mathcal{M}'_1 = b \sin z_{e1} = b$ and $u_{q1} = b\pi/8$. If we have instead $z_{e2} = 0$, $q_{e2} = a$ and $u_{q2} = 0$ whatever the uncertainty u_{z2}. This is obviously an incorrect result.

We now present three different alternative treatments, applicable to situations in which an output quantity is related to one input quantity through a nonlinear model relation and in which the LPU produces incorrect results.

3.5.1 Second-order Taylor series expansion

From the second-order Taylor series expansion of $\mathcal{M}(Z)$ about $Z = z_e$,

$$Q \approx \mathcal{M} + \mathcal{M}'(Z - z_e) + \tfrac{1}{2}\mathcal{M}''(Z - z_e)^2$$

where

$$\mathcal{M}'' \equiv \left.\frac{d^2 \mathcal{M}(Z)}{dZ^2}\right|_{Z=z_e}$$

we get

$$q_e = \mathrm{E}(Q) \approx \mathcal{M} + \tfrac{1}{2}\mathcal{M}'' u_z^2$$

68   *Evaluating the standard uncertainty*

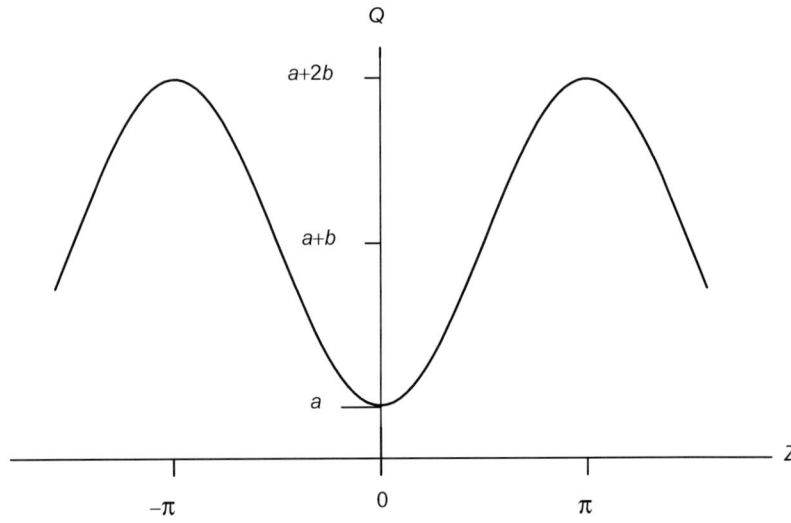

Figure 3.15. The model $Q = a + b(1 - \cos Z)$.

and

$$\begin{aligned}
u_q^2 &= V(Q) \\
&= E[Q - E(Q)]^2 \\
&\approx E\{\mathcal{M}'(Z - z_e) + \tfrac{1}{2}\mathcal{M}''[(Z - z_e)^2 - u_z^2]\}^2 \\
&= \mathcal{M}'^2 u_z^2 + \mathcal{M}'\mathcal{M}'' E(Z - z_e)^3 + \tfrac{1}{4}\mathcal{M}''^2 [E(Z - z_e)^4 - u_z^4]
\end{aligned}$$

where the last expression is obtained after some algebra. The third and fourth moments, $E(Z - z_e)^3$ and $E(Z - z_e)^4$, respectively, should be obtained from the pdf of Z.

Note that these formulae are also *approximate*. As with the LPU, they are valid only for values of Z close to z_e. Now, if $\mathcal{M}' = 0$,

$$u_q^2 \approx \tfrac{1}{4}\mathcal{M}''^2 [E(Z - z_e)^4 - u_z^4]. \tag{3.27}$$

Example. Consider again the model (3.26). This time let $Z = 2\pi t/t_p$, where t is the time at which the quantity Q is measured and t_p is the period of oscillation. If the time is not recorded, t should be considered as a random variable with a rectangular pdf defined over a domain of one period. Equivalently, we assume that the pdf associated with Z is rectangular over the interval $(-c, c)$, where $c = \pi$. This gives $z_e = 0$, $u_z^2 = c^2/3$, $\mathcal{M} = a$, $\mathcal{M}' = 0$, $\mathcal{M}'' = b$ and

$$E(Z)^4 = \frac{1}{2c}\int_{-c}^{c} z^4\, dz = \frac{c^4}{5}.$$

The special case of only one input quantity

Then
$$q_e \approx a + \frac{bc^2}{6} \tag{3.28}$$

and, from (3.27),
$$u_q^2 \approx \frac{(bc^2)^2}{45}. \tag{3.29}$$

However, these expressions are valid only for c close to zero, not for $c = \pi$.

3.5.2 Obtaining the expectation and variance directly

The second approach consists of obtaining directly (with no approximations) q_e as $E[\mathcal{M}(Z)]$ and u_q^2 as $V[\mathcal{M}(Z)]$.

Example. Consider once more the model (3.26). With Z as in the foregoing example, we have
$$q_e = a + b[1 - E(\cos Z)]$$

and
$$u_q^2 = b^2 V(\cos Z) = b^2 E(\cos^2 Z) - [bE(\cos Z)]^2$$

where
$$E(\cos Z) = \frac{1}{2c} \int_{-c}^{c} \cos z \, dz = \frac{1}{c} \sin c$$

and
$$E(\cos^2 Z) = \frac{1}{2c} \int_{-c}^{c} \cos^2 z \, dz = \frac{1}{2c} \sin c \cos c + \frac{1}{2}.$$

This gives
$$q_e = a + b\left(1 - \frac{1}{c} \sin c\right) \tag{3.30}$$

and
$$u_q^2 = b^2 \left(\frac{1}{2c} \sin c \cos c + \frac{1}{2} - \frac{1}{c^2} \sin^2 c \right). \tag{3.31}$$

It is left for the reader to verify that taking the expansions about zero of $\sin c$, $\sin c \cos c$ and $\sin^2 c$ to third, fifth and sixth order, respectively, one recovers (3.28) and (3.29).

Example. Let $Q = Z^2$, where Z is described by a pdf having expectation z_e and variance u_z^2. From (3.10) we find immediately
$$q_e = E(Q) = E(Z)^2 = z_e^2 + u_z^2$$

irrespective of the shape of $f(z)$. To obtain the variance u_q^2, however, the expectation $E(Z)^4$ needs to be computed.

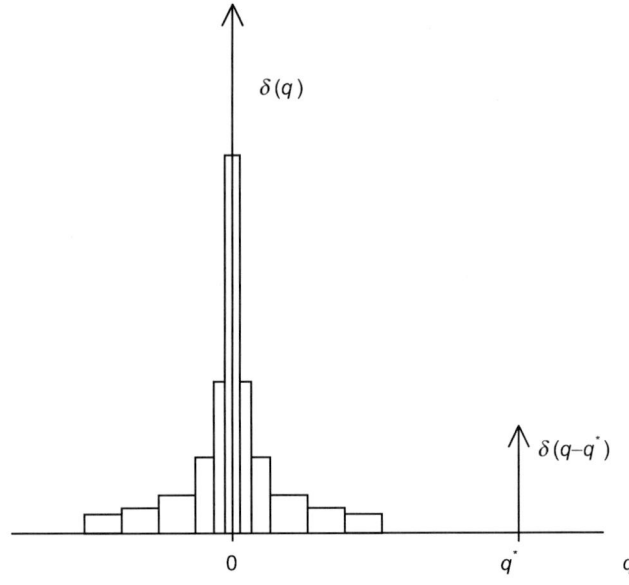

Figure 3.16. The delta function.

3.5.3 Obtaining the pdf: Dirac's delta function

The third, and more complete approach, consists of finding the pdf of Q from the pdf of Z. For this we must make a short detour to define a very useful mathematical tool, Dirac's *delta function*. This function can be thought of as the limit of a rectangular pdf whose width tends to zero. The normalization condition imposes its height to go to infinity. For this reason, the delta function is usually represented as a 'spike' over the location point (figure 3.16). The symbol for this function over the value $q = 0$ is $\delta(q)$. If the location point is q^*, we write either $\delta(q - q^*)$ or $\delta(q^* - q)$. From this definition, it is easy to see that

$$\int_{-\infty}^{\infty} \delta(q)\,dq = \int_{-\infty}^{\infty} \delta(q - q^*)\,dq = 1$$

whatever the value q^*. Moreover, it can be shown that, for any function $f(q)$,

$$\int_{-\infty}^{\infty} f(q)\delta(q - q^*)\,dq = f(q^*).$$

This is called the 'sifting' property of the delta function (see e.g. [19, p 653]).

Let us make use of (3.12) to write an expression for the joint pdf $f(q, z)$. This is

$$f(q, z) = f(q|z) f_Z(z)$$

where $f_Z(z)$ is the marginal pdf for Z and $f(q|z)$ is the conditional pdf for Q given that the value of Z is z (the subscript in $f_Z(z)$ is convenient to differentiate this pdf with the one associated with Q, $f_Q(q)$). To obtain $f(q|z)$, we note that when the value of Z is z, the model forces Q to assume the value $\mathcal{M}(z)$. This means that $f(q|z)$ becomes a delta function over the value $q = \mathcal{M}(z)$. Therefore

$$f(q,z) = f_Z(z)\delta[q - \mathcal{M}(z)].$$

The pdf $f_Q(q)$ is now found by integrating $f(q,z)$ over the domain D_Z of $f_Z(z)$ (note the similarity of this procedure with the one in section 3.3.3 to derive the log pdf). Thus

$$f_Q(q) = \int_{D_Z} f_Z(z)\,\delta[q - \mathcal{M}(z)]\,dz.$$

The integration is performed by means of the change of variables $y = \mathcal{M}(z)$. Then $z = \mathcal{M}^{-1}(y)$ and

$$dz = \frac{d\mathcal{M}^{-1}(y)}{dy}\,dy.$$

Therefore

$$f_Q(q) = \int_{D_Y} f_Z[\mathcal{M}^{-1}(y)]\left|\frac{d\mathcal{M}^{-1}(y)}{dy}\right|\delta(q - y)\,dy$$

where, since $Y = Q$, the domain D_Y coincides with the domain of $f_Q(q)$, and the absolute sign is used since the integrand, being a joint pdf, must be positive. From the sifting property we obtain, finally,

$$f_Q(q) = f_Z[\mathcal{M}^{-1}(q)]\left|\frac{d\mathcal{M}^{-1}(y)}{dy}\right|_{y=q}. \qquad (3.32)$$

This equation has to be used with care. It applies as if the function $\mathcal{M}(z)$ is single-valued, i.e. strictly monotonic (increasing or decreasing). Otherwise, its inverse is not properly defined. However, if a multiple-valued function is well behaved and can be written as a sum of single-valued functions $\mathcal{M}_i(z)$, as depicted in figure 3.17, equation (3.32) becomes

$$f_Q(q) = \sum_i f_Z[\mathcal{M}_i^{-1}(q)]\left|\frac{d\mathcal{M}_i^{-1}(y)}{dy}\right|_{y=q}.$$

This procedure will be generalized to the case of several input quantities in section 6.8.

Evaluating the standard uncertainty

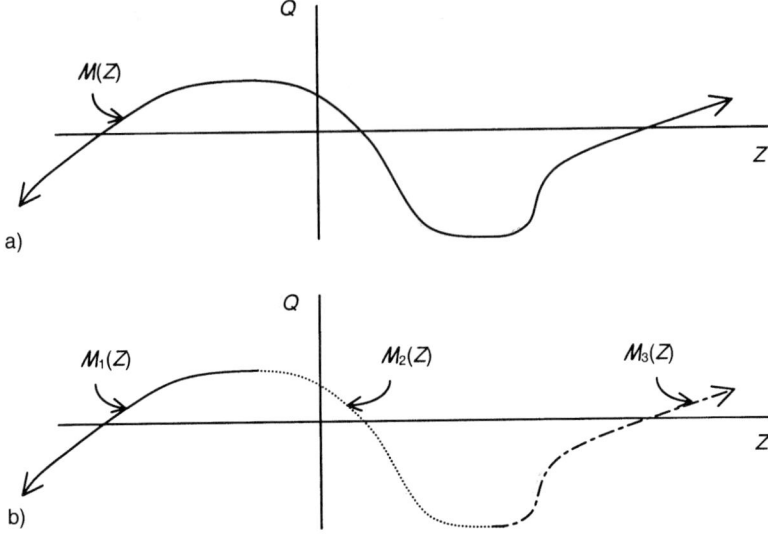

Figure 3.17. A multiple-valued function (*a*) expressed as a sum of single-valued functions (*b*).

Example. The transformation $Q = \ln Z$ is useful for expressing the value of a positive quantity Z in decibel units relative to some agreed-upon base value z° to which the value $q^\circ = \ln z^\circ = 0$ is assigned. We then have $\mathcal{M}^{-1}(y) = e^y$ and

$$f_Q(q) = f_Z(e^q)e^q.$$

For example, if $f_Z(z)$ is rectangular over the range $0 < a < z < b$, we get

$$f_Q(q) = \frac{e^q}{b-a}$$

for $\ln a < q < \ln b$. Integration yields

$$q_e = \frac{b(\ln b - 1) - a(\ln a - 1)}{b-a}.$$

and

$$u_q^2 = E(Q)^2 - q_e^2$$

where

$$E(Q)^2 = \frac{b[\ln^2 b + 2(1 - \ln b)] - a[\ln^2 a + 2(1 - \ln a)]}{b-a}.$$

Instead, the linearization of Q yields the approximate results

$$q_e \approx \ln z_e = \ln\left(\frac{a+b}{2}\right)$$

and

$$u_q \approx \frac{u_z}{z_e} = \frac{1}{\sqrt{3}}\frac{b-a}{b+a}$$

which are acceptable only for b very small or a very large.

Example. Consider the model (3.26) for the final time. Using figure 3.18 as a guide, we write

$$\mathcal{M}_1(z) = a + b(1 - \cos z) \qquad \text{for } -c \leq z \leq 0$$

and

$$\mathcal{M}_2(z) = a + b(1 - \cos z) \qquad \text{for } 0 \leq z \leq c.$$

Then, with $x = 1 - (y-a)/b$,

$$\mathcal{M}_i^{-1}(y) = \arccos x$$

$$\left|\frac{d\mathcal{M}_i^{-1}(y)}{dy}\right| = \frac{1}{b\sqrt{1-x^2}}$$

$$f_Z[\mathcal{M}_i^{-1}(q)] = \frac{1}{2c}$$

so that, finally,

$$f_Q(q) = \frac{1}{c\sqrt{(q-a)[2b-(q-a)]}} \qquad (3.33)$$

for $a < q < a + b(1 - \cos c)$.

Straightforward integration of (3.33) returns the results (3.30) and (3.31). This pdf may be called a 'distorted U', because for $c = \pi$ it acquires a symmetrical U shape about the expectation $q_e = a + b$ in the domain $(a, a+2b)$ with variance $u_q^2 = b^2/2$, see figure 3.19. (This last result is used without derivation in Clause H.1.3.4 of the *Guide*).

Example. Let $Q = Z^2$ and assume the pdf $f_Z(z)$ of Z is known. Setting $\mathcal{M}_1^{-1}(y) = \sqrt{y}$ and $\mathcal{M}_2^{-1}(y) = -\sqrt{y}$ yields the result

$$f_Q(q) = \frac{1}{2\sqrt{q}}[f_Z(\sqrt{q}) + f_Z(-\sqrt{q})]$$

defined for $q \geq 0$. As we have already seen, for this model we have

$$q_e = z_e^2 + u_z^2$$

74   Evaluating the standard uncertainty

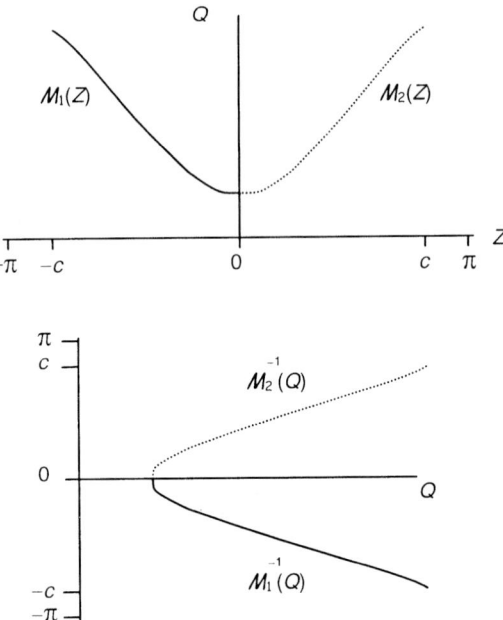

Figure 3.18. The model $Q = a(1 - \cos Z)$, $|z| \leq c \leq \pi$, and its inverse functions.

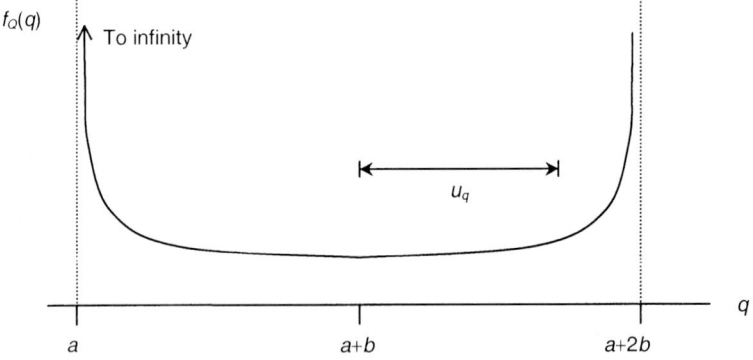

Figure 3.19. The 'U' pdf.

which is valid whatever the pdf $f_Z(z)$. If this is normal we get

$$f_Q(q) = \frac{1}{2u_z(2\pi q)^{1/2}} \left\{ \exp\left[-\frac{1}{2}\left(\frac{z_e - \sqrt{q}}{u_z}\right)^2\right] \right.$$
$$\left. + \exp\left[-\frac{1}{2}\left(\frac{z_e + \sqrt{q}}{u_z}\right)^2\right] \right\}.$$

Monte Carlo methods

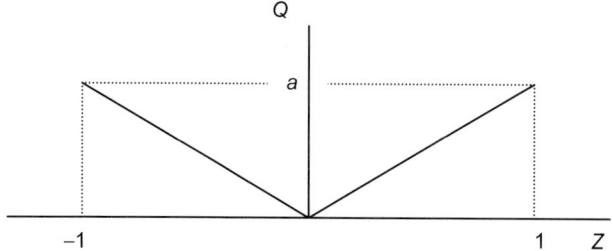

Figure 3.20. A linear model $Q = \mathcal{M}(Z)$ whose derivative is not defined at $z = 0$.

Example. Consider a quantity Q modelled as shown in figure 3.20, where a is a known constant. Even though the model is linear, its derivative at $z = 0$ is not defined. Therefore, if this point corresponds to the best estimate of the input quantity Z, neither the LPU nor the approaches in sections 3.5.1 and 3.5.2 can be applied. Assume the pdf of Z is rectangular, centred at $z_e = 0$ and of width 2. The procedure in this subsection then gives for Q a rectangular pdf centred at $q_e = a/2$ and of width a, from which $u_q = a/\sqrt{12}$ results.

3.6 Monte Carlo methods

We now describe an alternative to the LPU and to the procedure in section 3.5.3, called the *Monte Carlo* technique, for evaluating the uncertainty associated with the estimate of just one output quantity. The basic idea, which applies to linear as well as to nonlinear models, is extremely simple. Assume that in the model

$$Q = \mathcal{M}(Z_1, \ldots, Z_N)$$

all input quantities are described by known pdfs. A computer algorithm is set up to generate a sequence of values $\theta_1 = (z_1, \ldots, z_N)$, where each z_i is randomly generated according to the specific pdf for Z_i. The value q_1 corresponding to the sequence θ_1 is computed using the measurement model, and the process is repeated a large number of times. The outcome of this 'brute force' approach is a *frequency* distribution of values q_j, that approaches the *density* $f(q)$, whose mean and standard deviation can be numerically computed and then identified with the best estimate q_e and standard uncertainty u_q, respectively.

Note that this process is actually a simulation, because the quantity Q is not itself measured. Real measurements are only carried out in order to establish the pdfs of the Z_is. For details on the implementation of the Monte Carlo technique, see [6–8, 16, 17].

Example. $N = 1008$ random numbers z_i were generated according to

76   Evaluating the standard uncertainty

Table 3.1. Monte Carlo simulation of the model $Q = a+b(1-\cos Z)$ with $a = 0, b = 0.5$ and a rectangular pdf for Z in the interval $(-\pi, \pi)$.

j	q_j	n_j
1	0.05	217
2	0.15	94
3	0.25	66
4	0.35	95
5	0.45	65
6	0.55	57
7	0.65	61
8	0.75	97
9	0.85	77
10	0.95	179

a uniform density in the interval $(-\pi, \pi)$. For each z_i, the number

$$q_i = \frac{1 - \cos z_i}{2}$$

was calculated. This corresponds to the model (3.26) with $a = 0$ and $b = 0.5$. A count of the q_is falling in the intervals $(0, 0.1)$, $(0.1, 0.2)$, etc yielded the results in table 3.1, these are plotted in figure 3.21. The similarity between this figure and figure 3.19 is apparent. If the total n_j in each interval is assigned to the midpoint q_j of the interval, the numerical values

$$q_e = \frac{1}{N} \sum_{j=1}^{10} q_j n_j = 0.479$$

and

$$u_q = \left(\frac{1}{N} \sum_{j=1}^{10} q_j^2 n_j - q_e^2 \right)^{1/2} = 0.337$$

are obtained. These results should converge to those of the previous example ($q_e = a + b = 0.5$ and $u_q = b/\sqrt{2} = 0.250$) by increasing N and decreasing the width of the intervals.

3.7 Imported primary quantities

The input quantities to a measurement model are of two categories: those that are, in turn, submodelled as functions of other quantities; and those that are not modelled, at least in the current measurement task. The latter may be termed

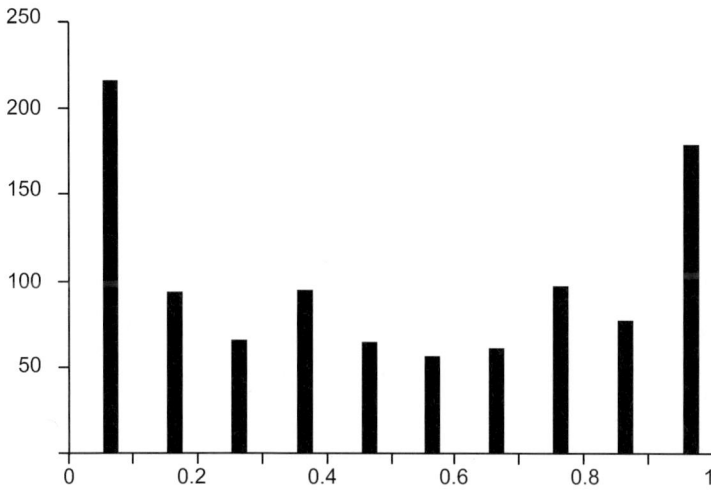

Figure 3.21. A plot of the results of a Monte Carlo simulation of the model $Q = a + b(1 - \cos Z)$ with $a = 0$, $b = 0.5$ and a rectangular pdf for Z in the interval $(-\pi, \pi)$.

primary quantities. By integrating the chain of submodels, it is possible to formulate the model in terms of primary input quantities only. For example, the uncertainty that we get for the modulus of elasticity would be the same whether we apply the model $E = \sigma/\varepsilon$ in figure 2.17 or $E = Ll_o/[ae(l-l_o)]$ in figure 2.18, as long as the same information is used in either case. Although this composition will seldom be done in practice, it serves to underscore the essential role played by the primary quantities: ultimately, every measurement model can be formulated in terms of quantities whose estimates and standard uncertainties are obtained directly from measurement or are given to the evaluator as the result of some other measurement procedure.

Let us consider first how to treat primary quantities that are *imported* from external sources. These are quantities whose best estimates are not obtained from the measurement task at hand, but from measurements carried out elsewhere (and wherein, they acted most probably not as primary quantities). A typical example is a quantity Q which is a function $\mathcal{M}(Z)$ of another quantity Z whose best estimate z_e is measured or otherwise known. However, instead of the function $\mathcal{M}$ being given analytically, as in equation (3.26), it is in the form of a table, say in a handbook of material properties. The best estimate of Q is then, simply, the value $q_e = \mathcal{M}(z_e)$ read from the table (figure 3.22).

In many cases the entries in the table will have been established empirically, on the basis of experiments carried out elsewhere. However rarely, if ever, will those values be given together with their corresponding uncertainties. In the absence of specific information on this matter, the user may use his/her judgement

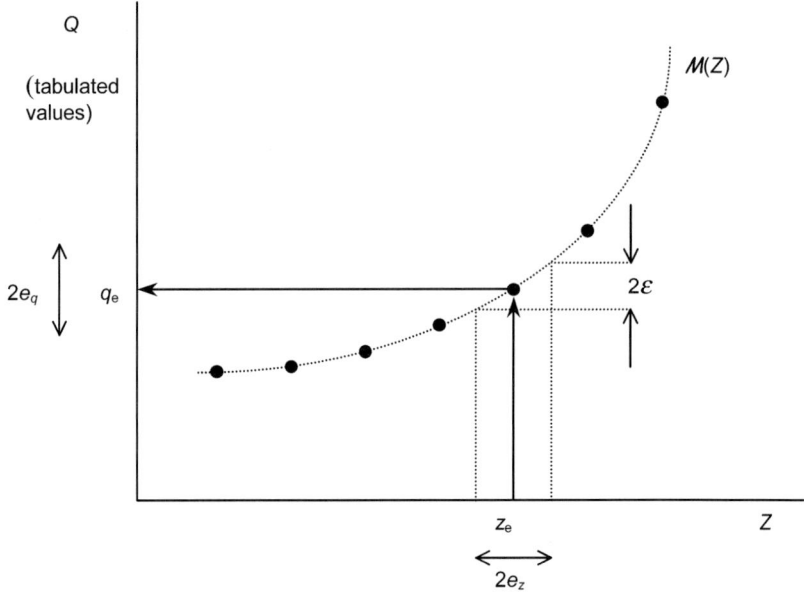

Figure 3.22. A quantity Q tabulated as a function of another quantity Z.

to set a maximum error e_q for the values in the table. To estimate a value of this parameter, one has to consider factors such as the presumed difficulty in the measurement of Q close to the value z_e of the quantity Z on which it depends, the year the table was constructed, the reliability of the source, the values of the same quantity tabulated in other handbooks, etc. Under the assumption that all values in the range $q_e \pm e_q$ are equally probable (a rectangular pdf), the uncertainty associated with q_e would then be given by equation (3.19):

$$u_q = \frac{e_q}{\sqrt{3}}.$$

An alternative approach is to assume that the measurement of Z is subject to a maximum error e_z. This leads to possible values for Q in the range $q_e \pm \varepsilon$, where

$$\varepsilon = \frac{|\mathcal{M}(z_e + e_z) - \mathcal{M}(z_e - e_z)|}{2}.$$

Since $\mathcal{M}$ is, in general, not given explicitly, the half-width ε will have to be established possibly by interpolation of the values in the table. If one assumes that the function $\mathcal{M}$ is not strongly nonlinear over the interval $z_e \pm e_z$, and uses again a rectangular pdf over this interval, the uncertainty associated with q_e would then be equal to

$$u_q = \frac{\varepsilon}{\sqrt{3}}.$$

Imported primary quantities 79

(Note that this equation is also obtained from the LPU by using $Q = \mathcal{M}(Z)$, $u_q = c_z u_z$, $c_z = \varepsilon/e_z$ and $u_z = e_z/\sqrt{3}$.)

Clearly, the first approach ignores the uncertainty of the input quantity Z, while the second approach does not take into account the uncertainty associated with the tabulated values.

Both uncertainty contributions are considered jointly if the model is written as $Q = X + Y$, where $X = \mathcal{M}(Z)$ represents the tabulated values (with $x_e = \mathcal{M}(z_e)$ and $u_x = e_q/\sqrt{3}$) and Y is a correction term (with $y_e = 0$ and $u_y = \varepsilon/\sqrt{3}$). From this model one gets

$$q_e = x_e = \mathcal{M}(z_e)$$

and

$$u_q^2 = u_x^2 + u_y^2 = \frac{e_q^2 + \varepsilon^2}{3}. \tag{3.34}$$

Example. It is desired to measure the flux of energy E passing from the burners of a steam power plant into the working fluid. From an energy balance of the boiler and superheater (figure 3.23), the model is

$$E = m(h_2 - h_1)$$

where m is the mass flow rate, h_1 is the enthalpy per unit mass of liquid water at the entrance to the boiler (pressure p_1 and temperature T_1), and h_2 is the enthalpy per unit mass of steam at the exit of the superheater (pressure p_2 and temperature T_2).

Measurement data are as follows: $m = 10.8$ kg s^{-1}, $p_1 = 500$ kPa, $T_1 = 30\,°\text{C}$ and $T_2 = 600\,°\text{C}$. Under the assumption of negligible pressure drops, we set $p_2 = p_1$. With these data we can now evaluate the enthalpies. Since the properties of a liquid depend very weakly on pressure, it is reasonable to assume that h_1 is equal to the enthalpy of saturated liquid water at T_1. Instead, the enthalpy of steam is a function of both temperature and pressure. Tables 3.2 and 3.3 show the handbook values. From these, we find $h_1 = 125.66$ kJ kg^{-1} and $h_2 = 3701.5$ kJ kg^{-1}. Therefore,

$$E = 10.8(3701.5 - 125.66)\ \text{kW} = 38\,619.072\ \text{kW}.$$

This estimate of energy flux is typical of what we would find as a student's answer in a thermodynamics quiz (in a sense it is correct, because to the student all data are 'perfectly known', including the tabulated enthalpy values). However, is it reasonable to state so many significant figures? This depends on the uncertainty associated with the estimates of the various input quantities. To assess $u(E)$ we use the LPU. Thus, from (3.25),

$$\frac{u^2(E)}{E^2} = \frac{u^2(m)}{m^2} + \frac{u^2(\Delta h)}{\Delta h^2}$$

Evaluating the standard uncertainty

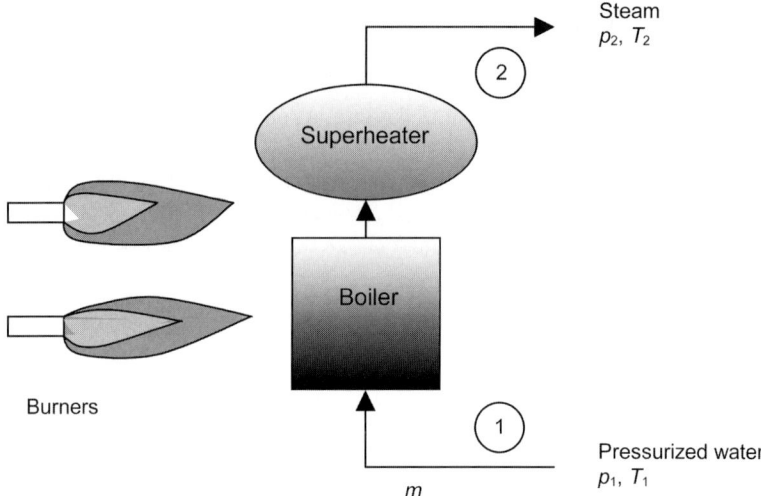

Figure 3.23. The steam generator of a power plant.

Table 3.2. Enthalpy of saturated liquid water.

$T/°C$	$h/(kJ\ kg^{-1})$
28	117.31
30	125.66
32	134.02

Table 3.3. Enthalpy of superheated steam.

	$h/(kJ\ kg^{-1})$		
p/kPa	$T = 500\ °C$	$T = 600\ °C$	$T = 700\ °C$
400	3484.9	3702.3	3926.4
500	3483.8	3701.5	3925.8
600	3482.7	3700.7	3925.1

or

$$u^2(E) = (h_2 - h_1)^2 u^2(m) + m^2[u^2(h_2) + u^2(h_1)]. \qquad (3.35)$$

Consider first $u(m)$. This uncertainty depends on how the mass flow rate was measured. Presumably, a device such as an orifice plaque, flow nozzle or venturi tube was used. These instruments are quite difficult to calibrate. In fact, they are often not calibrated at all. The user relies on empirical equations, obtained through rather careful experiments on geometrically similar devices, that give m as a function of the pressure drop through the device. Evidently, this is not the place to dwell on the details. Let us simply assume that the accuracy of the flow measuring instrument is 3 %. Therefore, the maximum error in the measurement of the mass flow rate is $10.8 \times 0.03 = 0.324$ kg s^{-1}. If a rectangular pdf is assumed, we get from equation (3.19):

$$u(m) = \frac{0.324}{\sqrt{3}} \text{ kg s}^{-1} = 0.187 \text{ kg s}^{-1}.$$

Since the tabulated enthalpy values do not carry an uncertainty statement, equation (3.34) will be used for the evaluation of $u(h_1)$ and $u(h_2)$. It will be assumed that the maximum error in the measurement of T_1 is 2 °C, and that of T_2 is 10 °C (the difficulties in the measurement of temperature increase the farther we are from ambient temperatures). As for the measurement of pressure, we will assume that it may be out by at most 50 kPa. Therefore, from table 3.2, $h_{1\min} = 117.31$ kJ kg^{-1} and $h_{1\max} = 134.02$ kJ kg^{-1}. Interpolating in table 3.3 we find $h_{2\min} = h(550 \text{ kPa}, 590\,°C) = 3679.315$ kJ kg^{-1} and $h_{2\max} = h(450 \text{ kPa}, 610\,°C) = 3724.320$ kJ kg^{-1}.

The estimated error bounds for the tabulated enthalpy values become

$$\varepsilon_{e1} = (134.02 - 117.31)/2 \text{ kJ kg}^{-1} = 8.355 \text{ kJ kg}^{-1}$$

and

$$\varepsilon_{e2} = (3724.320 - 3679.315)/2 \text{ kJ kg}^{-1} = 22.503 \text{ kJ kg}^{-1}.$$

It must now be assumed that the tabulated values themselves may be in error. Using $e_1 = 1$ kJ kg^{-1} and $e_2 = 20$ kJ kg^{-1}, from (3.34) we get $u(h_1) = 4.86$ kJ kg^{-1} and $u(h_2) = 17.38$ kJ kg^{-1}. Substitution of these values into (3.35) we find, finally, $u(E) = 697$ kW. According to the *Guide*, the standard uncertainty should be reported with just two significant figures. We then write $u(E) = 0.70$ MW and, to be consistent, the estimate of energy flux is reported as $E = 38.62$ MW.

3.8 Directly measured primary quantities: type A analysis

We turn now to the other type of primary quantities: those that are measured directly using a measuring instrument or measuring system. The distinction between these is only of complexity. The former is usually a single apparatus, while the latter involves a set of instruments and other equipment assembled to carry out measurements. In any case, the final result of applying an instrument or system to a measurand Q is a gauged or calculated indication G about the value of the measurand. In general, the relation between Q and G will involve an additional correction quantity C or F (and perhaps a quantity related to the resolution of the instrument, see section 3.9). From the two alternative models

$$Q = G + C \qquad (3.36)$$

or

$$Q = FG \qquad (3.37)$$

we get

$$q_e = g_e + c_e$$

and

$$u_q^2 = u_g^2 + u_c^2$$

or

$$q_e = f_e g_e$$

and

$$u_q^2 = g_e^2 u_f^2 + f_e^2 u_g^2$$

where the LPU has been used under the assumption that G and C (or F) are independent.

Models for the correction quantities will be presented in sections 3.10 and 3.11. The model for the quantity G depends on the measuring system. If this quantity is determined by means of a single instrument, there will usually be no model at all. It will then be necessary to rely on the characteristics of the instrument and on the information it renders about the measurand. In this section we address the case where the resolution of the indicating device does not affect the indications. Digital display devices are considered in section 3.9.

Assume that Q is a non-varying, stable quantity that is directly and repeatedly measured under repeatability conditions with an instrument that shows negligible drift during the observation period. This produces a series of *independent* indications $\theta = (g_1, \ldots, g_n)$, $n \geq 2$. The procedure recommended in the *Guide* to deal with this sort of information is founded on *sampling theory*, or frequency-based statistics, and does not involve the calculation of the expectation and variance of some pdf. Instead, the procedure relies on assuming the existence of an underlying *frequency distribution* $f(g)$, where g represents a generic element that belongs to an imagined infinite *population* of gauged values. While the expectation and the variance of $f(g)$ are unknown, their values may be *estimated* from the information provided by the series θ, which is termed a *random sample* from the population. It is thus seen that the theoretical basis of this type of calculations is completely different from that explained so far. For this reason, the *Guide* refers to this as 'type A' analysis, TA for short. (All 'other

Directly measured primary quantities: type A analysis

methods' of uncertainty evaluation are lumped therein under the category 'type B'.)

The procedure is best understood by considering a series of independent random variables $\Theta = (G_1, \ldots, G_n)$, where the distributions of all the G_is are equal to $f(g)$. The ith indication g_i is taken as the particular value assumed by the corresponding variable G_i. If another random sample was drawn, a different set of values would be taken up by the series Θ. Therefore, any given function of the variables in Θ, called a *statistic*, will itself be a random variable. For our purposes, the most important statistics are the *sample mean*

$$\overline{G} = \frac{1}{n} \sum_{i=1}^{n} G_i$$

and the *sample variance*

$$S^2 = \frac{1}{n-1} \sum_{i=1}^{n} (G_i - \overline{G})^2.$$

After the measuring sequence has been carried out, these statistics assume the values

$$\overline{g} = \frac{1}{n} \sum_{i=1}^{n} g_i$$

and

$$s^2 = \frac{1}{n-1} \sum_{i=1}^{n} (g_i - \overline{g})^2$$

respectively.

Now let μ be the expectation of $f(g)$ and σ^2 its variance. It can be shown that, irrespective of the form of this distribution,

$$E(\overline{G}) = \mu \tag{3.38}$$

and

$$E(S)^2 = \sigma^2. \tag{3.39}$$

These formulae are proved, e.g., in [38, pp 270 and 274].

According to (3.38), $\overline{g}$ is an unbiased estimate of the expectation μ. This estimate is taken as the best estimate of G and, therefore,

$$g_e = \overline{g}. \tag{3.40}$$

Similarly, (3.39) states that s^2 is an unbiased estimate of the variance σ^2 or, equivalently, that s estimates the standard deviation σ. Since the latter is the typical spread of an *individual* member of the population, a value g_i drawn at random from the sample is expected, on average, to deviate from $\overline{g}$ by the amount

s. Therefore, s is taken as a quantitative indication of the *precision* of the random sample θ (the higher the value s is, the smaller the precision becomes). Now, it would be reasonable to state s to be the standard uncertainty associated with *one* of the values g_i if *that value* was given as the estimate of the quantity. But since we use $\bar{g}$, an estimate of the typical spread of $\overline{G}$ is needed. For this we use the fact that

$$V(\overline{G}) = \frac{\sigma^2}{n}$$

and therefore the standard deviation $S(\overline{G})$ is estimated by

$$u_g = \frac{s}{\sqrt{n}} \qquad (3.41)$$

which is identified with the standard uncertainty associated with g_e. (In section 2.7 it was mentioned that the usual expression 'uncertainty estimation' is not proper, and this is also true in this case. What one does with the TA analysis is to estimate a *standard deviation* and to *identify* this estimate with the standard uncertainty.)

Example. A quantity G is measured repeatedly 10 times. Numeric values are

$$7.489$$
$$7.503$$
$$7.433$$
$$7.549$$
$$7.526$$
$$7.396$$
$$7.543$$
$$7.509$$
$$7.504$$
$$7.383$$

and the unit is mm. These values give $g_e = \bar{g} = 7.484$ mm and $u_g = 0.019$ mm. The precision is $s = 0.059$ mm.

Note that u_g may turn out to be zero. Because of the models (3.36) and (3.37), this does not necessarily mean that the measured quantity Q is perfectly known. Now if u_g does not vanish, it is because at least one of the indications g_i is different from the rest. Unless the instrument is defective, this happens, in general, because of the action of influence quantities on repeatable measurements. Thus, a standard uncertainty evaluated as per equation (3.41) may be termed a *repeatability* component of uncertainty.

The *Guide* mentions that the TA method of evaluation should be used if n is 'large enough', say greater than 10. For smaller samples, there are three alternative approaches. The first applies when there is available prior information on the instrument's behaviour. This may take the form of a sample variance s'^2 obtained from n' repeated measurements of a similar measurand under similar conditions of measurement, where n' is large. Then, instead of calculating the precision s from the actual (small) sample, we use instead s' and write

$$u_g = \frac{s'}{\sqrt{n}} \qquad (3.42)$$

which is valid even for $n = 1$. The *Guide* refers to s'^2 as a 'pooled' estimate of variance, and denotes it as s_p^2. In this approach, the applicable *degrees of freedom* are $n' - 1$ (see Clauses 4.2.4 and F.2.4.1 of the *Guide*, see also section 4.5).

The second approach applies if s' is not available. One can then use for G a rectangular pdf centred about the average of the (few) measured values, the width of which has to be determined from whatever information about the measurand and the instrument characteristics is available. If no such information exists, the rectangular pdf can be centred at the midpoint between $g_{i,\max}$ and $g_{i,\min}$, with a width equal to the difference between these values.

Example. A gauge block of unknown length l is calibrated using a mechanical comparator. The measurement standard is another gauge block whose known conventional true value is l_s. The difference $d = l - l_s$ is measured $n = 5$ times; the results are (215, 212, 211, 213, 213) nm. From these values the best estimate of l is the average 212.8 nm and the standard uncertainty computed from (3.41) is 0.66 nm with a precision $s = 1.48$ nm. Suppose that, from $n' = 25$ similar comparison measurements carried out using another set of blocks, a precision $s' = 1.66$ nm was computed. The best estimate is again 212.8 nm, but using (3.42) the standard uncertainty is assessed to be 0.74 nm. Finally, if a rectangular density is assumed over the maximum and minimum of the measured values, the best estimate becomes $(215 + 211)/2$ nm $= 213.0$ nm with a standard uncertainty equal to $(215 - 211)/\sqrt{12}$ nm $= 1.2$ nm.

The third approach is based on Bayesian statistics. It will studied in section 6.5.

3.9 Digital display devices

If, in the models $Q = G + C$ or $Q = FG$, it turns out that $c_e = 0$ or $f_e = 1$ with negligible uncertainty, one would use the model $Q = G$, thereby expressing

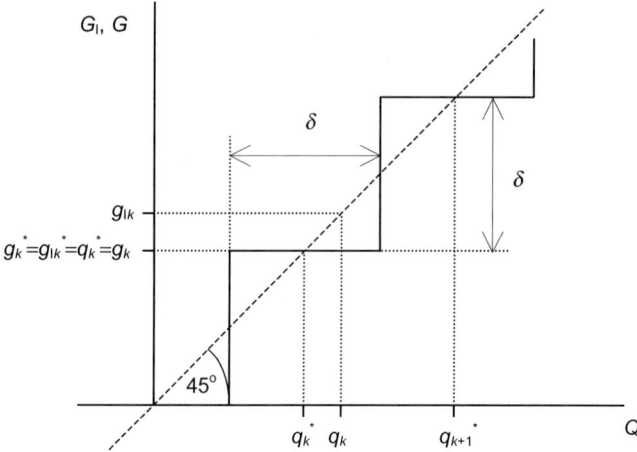

Figure 3.24. The indications G of a digital display device having a digital step equal to δ.

the fact that the only information required to infer the value of the measurand Q concerns the indications G. Now assume that data about the latter consist of a series of observations obtained under the same conditions of measurement. If the indicating device of the instrument does not have enough resolution to detect the possible random scatter caused by influence quantities, all indications would be the same and a vanishing uncertainty would result from the TA analysis for both G and Q. This would imply that the measured quantity Q is perfectly known, which is obviously incorrect.

To analyse how the digital resolution of the instrument's display device affects the uncertainty associated with the estimate of Q, consider figure 3.24. The abscissae represent the possible values of the measurand and the ordinates represent the indications G_I and G. The former correspond to an ideal instrument that is not limited by resolution, they are found from the dotted line at 45° (i.e. $G_I = Q$) if no correction quantities are needed. The indications G of the real instrument are found from the full ladder-like line. Thus, if Q takes the value $q_k^* = k\delta$, where k is an integer and δ is the digital step, both the ideal and actual instrument would produce the indication $g_k^* = g_{Ik}^* = q_k^*$. If, instead, Q takes the value q_k, with $|q_k - q_k^*| < \delta/2$, the ideal instrument would produce the value $g_{Ik} = q_k$ while the real instrument would give again the indication $g_k = g_k^* = q_k^*$.

Therefore, we can model this problem by defining a quantity R equal to the difference $G_I - G = Q - G$, such that $Q = G + R$ [27]. Since it is known that the value of R cannot exceed half the digital step, we take for this quantity a rectangular pdf centred at zero and of width δ. This gives

$$q_e = g_e + r_e = g_e$$

Table 3.4. A quantity Q measured repeatedly with two different instruments.

	Instrument 1 $\delta_1 = 0.001$	Instrument 2 $\delta_2 = 0.1$
g_1	7.489	7.5
g_2	7.503	7.5
g_3	7.433	7.4
g_4	7.549	7.5
g_5	7.526	7.5
g_6	7.396	7.4
g_7	7.543	7.5
g_8	7.509	7.5
g_9	7.504	7.5
g_{10}	7.383	7.4
q_e	7.4835	7.4700
u_g	0.0187	0.0153
u_r	0.0003	0.0289
u_q	0.0187	0.0327

and

$$u_q^2 = u_g^2 + u_r^2 = u_g^2 + \frac{\delta^2}{12}$$

where g_e and u_g are obtained following the TA procedure.

Example. A quantity Q is repeatedly measured with an instrument whose digital step is $\delta_1 = 0.001$ (arbitrary units). The data and results are shown in the column labelled 'Instrument 1' of table 3.4. This table also gives the data and results that would be obtained if another instrument with $\delta_2 = 0.1$ was used. For the first instrument we get $u_q = u_g$, meaning that the contribution of the uncertainty $u_r = \delta/\sqrt{12}$ is negligible. This does not happen for the second instrument.

The model $Q = G + R$ can also be applied to analogue display devices with δ equal to the difference between the values corresponding to two successive scale marks. However, in this case a better choice for the pdf of R might be the logarithmic one (section 3.3.3) centred at $r_e = 0$ with parameters e_e and ε that depend on factors such as the width and separation of the marks, the visual acuity of the observer, the influence quantities and even the nature of the measurand itself.

An alternative procedure to deal with the case of analogue-to-digital conversion errors, based on Bayesian statistics, is presented in section 6.5.3.

3.10 Correction quantities: systematic effects

We now turn to the correction quantities C and F in the models (3.36) and (3.37). Let us begin by considering the following definitions in the VIM:

Correction: value added algebraically to the uncorrected result of a measurement to compensate for systematic error.

Correction factor: numerical factor by which the uncorrected result of a measurement is multiplied to compensate for systematic error.

Both definitions depend on the concept of *systematic error*: 'the mean that would result from a very large number of measurements of the same measurand carried out under repeatability conditions minus a true value of a measurand'. As discussed in section 2.5, this definition is not operational, because true values cannot be known. However, the intuitive meaning of systematic error is clear: it corresponds to the difference between the actual value we get from repeated measurements and the value we 'should have obtained' if each measurement had been 'perfect'.

Despite the widespread usage of the term 'systematic error', we find it preferable to use instead the related concept—not defined in the VIM—of *systematic effect*: an influence that leads to an identifiable and fixed but uncertain bias in the measurement result with respect to the value of the measurand. Examples are: imperfect alignment, thermal expansion, lead resistance, load impedance, thermal voltages, leakage currents, etc. The adjective 'identifiable' is used here as a synonym for *modellable*: the bias can be made to enter into the measurement models (3.36) or (3.37) through the quantities C or F, intended to compensate for the action of the systematic influence. Evidently, C and F correspond basically to the VIM's definitions for correction and correction factor, respectively. However, these terms are here defined as *quantities*: through adequate modelling and measurement, it should be possible to obtain their best estimates and associated uncertainties.

The estimates of the correction quantities may be such that they have no effect on the estimated value of the measurand (i.e. it may be that $c_e = 0$ or $f_e = 1$). However, the uncertainties associated with the correction estimates will cause an increase in the uncertainty associated with the estimate of the measurand.

Example. It is desired to measure the volume V_f° contained in a glass flask up to an engraved mark at a specified reference temperature T°, say 20 °C (figure 3.25). For this, the following procedure is used:
- The flask is washed and dried thoroughly and placed in a room maintained at $T_a \approx T^\circ$.
- The ambient temperature T_a and pressure p_a are measured.
- A sample of distilled water is placed in a separate container. The water temperature T_1 is measured and the procedure is continued if the reading is close to T°.

Correction quantities: systematic effects

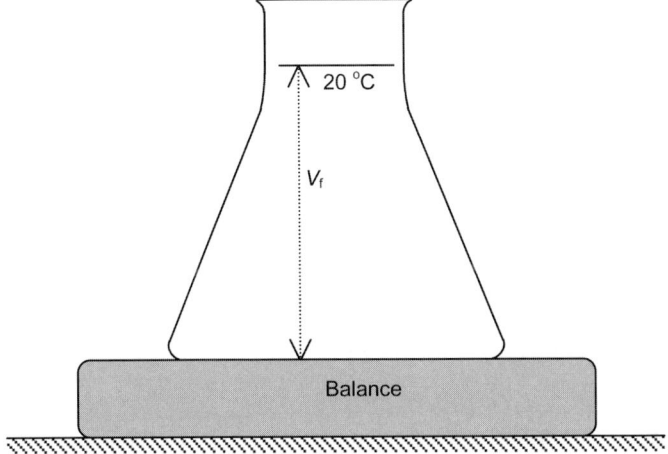

Figure 3.25. The measurement of the volume of a flask.

- The balance with the empty flask is zeroed.
- The flask is filled up to the mark with water from the container.
- The flask is weighed to determine the indicated mass of water, m_{Iw}.
- The flask is emptied and the weighing is repeated several times.
- The water density ρ_w is measured with a densitometer.
- The water temperature T_2 in the container is measured to check that it has not drifted away from $T°$.

If systematic effects are disregarded, the measurement model is simply

$$V_f° = V_w$$

where V_w is the volume of water in the flask. In turn, for V_w we use the model

$$V_w = \frac{m_{Iw}}{\rho_w}.$$

However, depending on the desired degree of accuracy, systematic effects caused by temperature and buoyancy should be considered. This is done as follows.

Let T be temperature of the flask at the time of measurement. The relation between the volume at this temperature and that at $T°$ can be modelled as

$$V_f° = V_f(T)[1 - \alpha(T - T°)]$$

where α is the volumetric coefficient of thermal expansion of the flask's material. Therefore, the quantity

$$F_T = 1 - \alpha(T - T°)$$

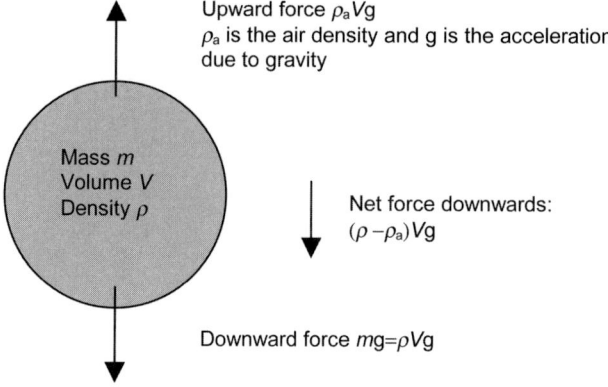

Figure 3.26. Buoyancy as a systematic effect.

becomes a thermal correction factor and we have

$$V_f^\circ = F_T V_f(T)$$

where

$$V_f(T) = \frac{m_w}{\rho_w}$$

and where m_w is the actual mass of water in the flask. In turn, the temperature T may be modelled as

$$T = \frac{T_1 + T_2}{2}.$$

The second systematic effect has to do with the fact that the balance reacts not to mass but to the downward force acting on its plate. This force is the resultant of the opposing actions of gravity and air buoyancy (figure 3.26). Since the balance was zeroed with the empty flask, its indications are proportional to

$$(\rho_w - \rho_a) V_w g$$

where ρ_a is the air density and g is the acceleration of gravity. Instead, the corrected indications should be proportional to

$$\rho_w V_w g.$$

The model for the actual mass of water in terms of the indicated mass of water then becomes

$$m_w = \frac{m_{Iw}}{F_a}$$

where
$$F_a = 1 - \frac{\rho_a}{\rho_w}$$
is a buoyancy correction factor (or, more strictly, the inverse of a correction factor, according to the VIM's definition).

The *corrected* measurement model then becomes

$$V_f^\circ = \frac{m_{Iw} F_T}{\rho_w F_a}. \tag{3.43}$$

To evaluate F_a we need a model for the air density. This is given by the constitutive relation for ideal gases:

$$\rho_a = \frac{p_a}{R_a T_a}$$

where R_a is the ideal gas constant for air and p_a and T_a are the ambient air pressure and temperature at the time of weighing. (A correction factor might also be used on the ideal gas model. However, for air at atmospheric pressures, this factor is nearly equal to 1, with negligible uncertainty.)

The primary input quantities to the model are then m_{Iw}, α, T_1, T_2, ρ_w, p_a and T_a. Of these, α is imported and the others are measured directly. The estimates and standard uncertainties of all these quantities may be evaluated as explained in sections 3.7–3.9, possibly in combination with calibration corrections, to which the next section is devoted.

To evaluate the uncertainty associated with the estimate for the volume of the flask, we might think of using (3.25). We would then obtain

$$\frac{u^2(V_f^\circ)}{V_f^{\circ 2}} = \frac{u^2(m_{Iw})}{m_{Iw}^2} + \frac{u^2(\rho_w)}{\rho_w^2} + \frac{u^2(F_T)}{F_T^2} + \frac{u^2(F_a)}{F_a^2}.$$

However, this expression is incorrect, because it does not take into account the obvious correlation between ρ_w and F_a. This can be eliminated by writing the model in terms of the independent input quantities. Thus, from

$$V_f^\circ = \frac{m_{Iw} F_T}{\Delta \rho}$$

where

$$\Delta \rho = \rho_w - \rho_a$$

we get

$$\frac{u^2(V_f^\circ)}{V_f^{\circ 2}} = \frac{u^2(m_{Iw})}{m_{Iw}^2} + \frac{u^2(F_T)}{F_T^2} + \frac{u^2(\Delta \rho)}{\Delta \rho^2}$$

Table 3.5. Data for the determination of the conventional true value of a flask's volume.

Quantity	Estimate	Std uncertainty	Unit
m_{Iw}	99.8382	0.0032	g
α	18	5	$(\text{MK})^{-1}$
$T_1 = T_2$	20.1	0.014	°C
ρ_{w}	997.820	0.059	g L^{-1}
p_{a}	101	1	kPa
T_{a}	20.5	0.03	°C

with

$$\frac{u^2(F_T)}{F_T^2} = \frac{u^2(\alpha)}{\alpha^2} + \frac{u^2(T)}{T^2}$$

$$u^2(\Delta_\rho) = u^2(\rho_{\text{w}}) + u^2(\rho_{\text{a}})$$

and

$$\frac{u^2(\rho_{\text{a}})}{\rho_{\text{a}}^2} = \frac{u^2(p_{\text{a}})}{p_{\text{a}}^2} + \frac{u^2(T_{\text{a}})}{T_{\text{a}}^2}.$$

As a numerical example, consider the data in table 3.5. These pertain to a flask of 0.1 L nominal volume. With $T° = 20\,°\text{C}$ and $R_{\text{a}} = 0.287\,08$ kJ (kg K)$^{-1}$, results are: $\rho_{\text{a}} = 1.198$ kg m^{-3}, $F_{\text{a}} = 0.998\,799$, $F_T = 0.999\,998$ and $V_{\text{f}}° = 0.1$ L with a standard uncertainty $u(V_{\text{f}}°) = 28$ mL.

3.11 Correction quantities: calibration

A calibration is, in a sense, a procedure to 'measure' a correction or a correction factor. The VIM's definition of calibration is

> *the set of operations that establish, under specified conditions, the relationship between values of quantities indicated by a measuring instrument or measuring system, or values represented by a material measure or a reference material, and the corresponding values realized by standards.*

In simpler terms, a calibration is intended to relate the value or values of a *calibrand* with the value or values of one or more measurement standards. The calibrand may be

- a measuring instrument or system,

- a material measure or
- a reference material.

In section 3.8 we have distinguished between measuring instruments and measuring systems. For their part, material measures and reference materials are not intended to carry out measurements, but to *reproduce* (i.e. make available) the value of a given quantity. Examples of material measures are: gauge blocks, reference electrical resistors, fixed temperature points, standard masses, etc. The flask of the example in section 3.10 may be considered as a material measure to reproduce a volume. Reference materials are similar to material measures, but the quantities these materials reproduce are properties such as pH, viscosity, chemical composition, heat capacity and the like. In analytical and chemical metrology, the importance of reference materials is paramount.

In turn, the distinction between a calibrand and a measurement standard is only that of occasion: once a calibrand has been calibrated, it might serve as a standard in the calibration of another calibrand. Thus a standard may be a measuring instrument, a measuring system, a material measure or a reference material. Most (calibrated) material measures and (certified) reference materials are used in practice as standards to calibrate measuring instruments or systems.

Let us now consider 'the set of operations that establish the relationship between the value of the calibrand and the value of the standard'. These depend basically on the calibrand, because the standard is chosen according to the nature of the former. For a material measure or reference material, the calibration is intended to determine the conventional true value realized by the object or substance (or its difference with respect to the nominal value). Instead, for a measuring instrument or system, the purpose of calibration may be: (i) to establish the position of the scale marks (a process the VIM defines as *gauging*) or (ii) to establish the corrections or correction factors that should be applied to the indications when the scale is already in place. (The term 'scale marks' should be interpreted here as encompassing those physically engraved or painted in the display device—usually of analogue type—as well as those that are rendered visible at the time the indication is displayed, as is the case for most digital display devices). In short, through a calibration one determines the conventional true value of a material measure or reference material, or establishes what an instrument or system 'should indicate'.

In practice, a wide variety of calibrands and standards is encountered. Nevertheless, it is possible to give a general idea of the sort of operations performed during a calibration. These are carried out in facilities known generically as *calibration laboratories* and these are described next. Reference [10] gives several examples of calibration procedures and of their evaluation.

3.11.1 Calibration of material measures

Consider, first, material measures and reference materials. For these types of calibrands it may be possible to evaluate the estimate and associated standard

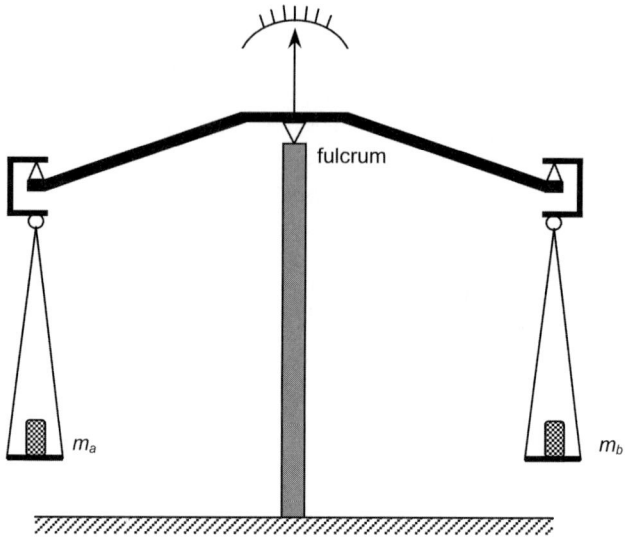

Figure 3.27. The equal-arm beam balance.

uncertainty of the test quantity T it realizes by means of a measuring instrument or a measuring system. For the latter to be regarded as the calibration standard, it should provide the traceable indications needed to evaluate the estimates and uncertainties of the primary input quantities to the model for T. The determination of the flask volume V_f modelled by (3.43) is an example of such a calibration method. An alternative procedure consists of using a *comparator* to evaluate T on the basis of the value of a calibrated standard S. The following are two examples of comparison measurements.

> **Example.** Consider the equal-arm beam balance. (This the international symbol of justice, figure 3.27, and is perhaps the oldest comparator developed by man.) The beam is suspended at its exact centre on a knife-edge. The point of support is called the fulcrum. Two pans of equal weight are suspended from the beam, one at each end, at points equidistant from the fulcrum. The beam will be in equilibrium, as evidenced by the vertical position of the pointer at the top, when the masses m_a and m_b that are placed over each pan (corrected for the effects of buoyancy if their volumes are different) are equal.
>
> In calibration, one of the pans is loaded with the test specimen T, the other pan is loaded with the *tare weight* whose mass is close to that of T, but which does not need to be known accurately. The rest position P_T of the pointer is recorded. The test specimen is then substituted by the standard S and position P_S results. The difference $D = P_T - P_S$, which can be positive or negative, is converted into

Correction quantities: calibration   95

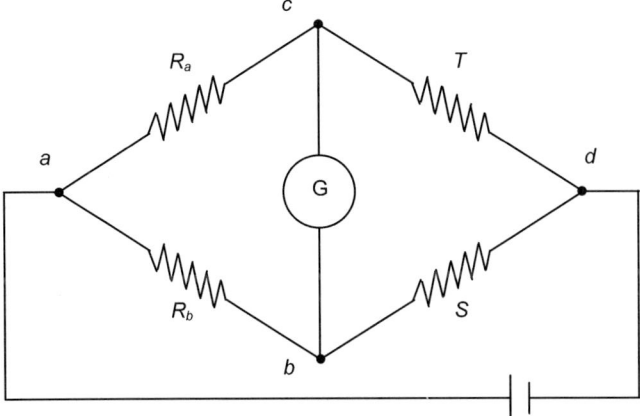

Figure 3.28. The Wheatstone bridge.

a mass value through division by the balance's sensitivity. From the model $T = S + D$ we get

$$t_e = s_e + d_e$$

and

$$u_t^2 = u_s^2 + u_d^2.$$

Example. The Wheatstone bridge (figure 3.28) is the electrical equivalent to the equal-arm beam balance. The circuit consists of four resistors connected in series–parallel arrangement with a voltage source applied at opposite corners (a and d) and a galvanometer G connected across the other two corners (b and c). Resistors R_a and R_b are selected for a fixed ratio $R = R_a/R_b$ and the standard S is adjusted until no potential difference exists between points c and d, as evidenced by zero deflection of the galvanometer. This null condition is obtained when the resistors satisfy

$$T = SR$$

from which model we get

$$t_e = s_e r_e$$

and

$$\frac{u_t^2}{t_e^2} = \frac{u_s^2}{s_e^2} + \frac{u_r^2}{r_e^2}.$$

Further considerations with regards to the analysis of comparison measurements are given in section 5.4 and in section 5.8.3.

Whatever the procedure for measuring T, the calibration certificate can give either or both: (i) the best estimate t_e and the standard uncertainty u_t, or (ii) the value and uncertainty associated with the ratio or the difference between T and the *nominal value* t_n of the calibrand. (The nominal value is a rounded or approximate value of interest as a guide to the user of the calibrand and, as such, it carries no uncertainty.) The difference $t_e - t_n$ is defined in the VIM as the *deviation* of the material measure.

3.11.2 Calibration of measuring instruments

Measuring instruments and measuring systems are usually calibrated by 'measuring' a standard S to obtain the estimate i_e and the standard uncertainty u_i for the indications I during calibration.

From the alternative models

$$C = S - I$$

or

$$F = \frac{S}{I}$$

we get

$$c_e = s_e - i_e$$

and

$$u_c^2 = u_s^2 + u_i^2$$

or

$$f_e = \frac{s_e}{i_e}$$

and

$$\frac{u_f^2}{f_e^2} = \frac{u_s^2}{s_e^2} + \frac{u_i^2}{i_e^2}.$$

According to the chosen model, the estimate and standard uncertainty of the calibration corrections C or F are reported in the calibration certificate of the instrument (more often, however, instrument calibration certificates will state the 'error of indication' $-c_e$). In general, the quantities C or F will enter as input quantities to the corrections in models (3.36) and (3.37). This is because the latter may include other additive or multiplicative corrections for systematic effects that take place during the measurement of Q.

3.11.3 Calibration curve

Most measuring instruments are intended to be used over a certain range of values of the measurands to which they might be applied. For this reason, in practice the calibration operations should be carried out at various points

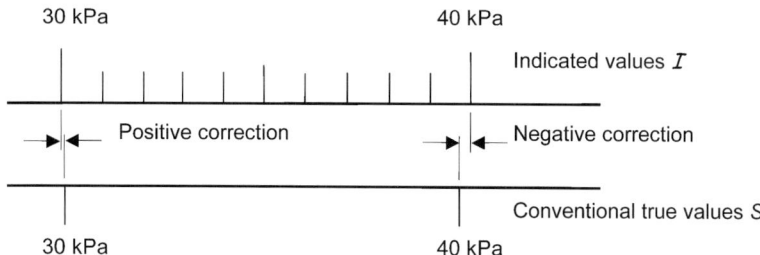

Figure 3.29. Calibration of a manometer at 10 kPa intervals.

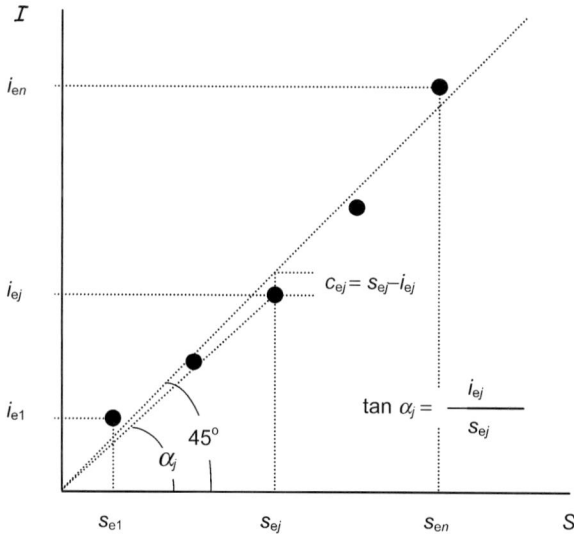

Figure 3.30. A set of discrete calibration points covering the measuring range of a calibrated instrument.

that cover the measuring range, see figures 3.29 and 3.30. The calibration is performed with standards $S_1, \ldots, S_n$, to which indications $I_1, \ldots, I_n$ correspond. If the instrument was ideal, i.e. it had no errors, the estimated points (s_{ej}, i_{ej}), $j = 1, \ldots, n$ would fall around a straight line bisecting the axes. The vertical distances of the calibration points from this line are the estimates of the calibration corrections $c_{ej} = s_{ej} - i_{ej}$. The angles α_j of the lines joining the origin and the calibration points are the arc cotangents of the estimates of the correction factors $f_{ej} = s_{ej}/i_{ej}$. Uncertainties (u_{sj}, u_{ij}) and, possibly, a number of covariances corresponds to each pair (s_{ej}, i_{ej}).

Since the estimates c_{ej} or f_{ej}, and their uncertainties, apply only to the specific point where the calibration was performed, it may be preferable to report the results of the calibration in the form of a continuos interpolation curve that

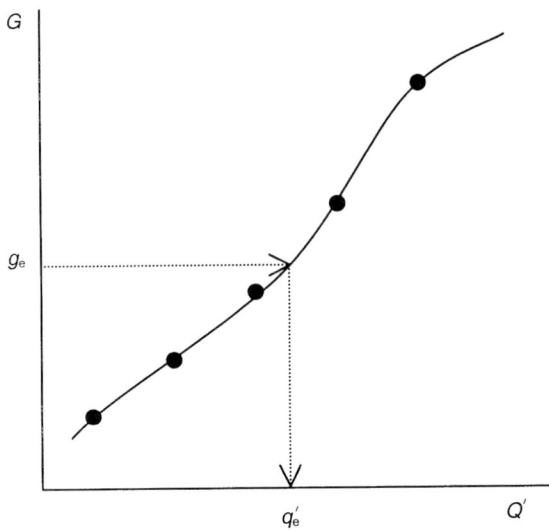

Figure 3.31. A calibration curve.

passes over or near the individual calibration points. Such a curve is depicted in figure 3.31. Note that the labels of the coordinate axes have been changed with respect to those in figure 3.30: in practice, a value g_e for the uncorrected indications G is obtained by direct measurement and the estimated value q'_e of the quantity Q' is read from the curve, where Q' represents the uncorrected measurand (other corrections or correction factors may have to be applied to Q' in order to reach the measurand Q).

The calibration curve is, in general, given in terms of a number of parameters A_k. These are interpreted as quantities with estimates a_{ek}, uncertainties u_{ak} and mutual uncertainties $u_{ak,aj}$. Typically, calibration curves are polynomials, i.e.

$$G = \sum_{k=0}^{N} A_k Q'^k$$

where the degree N is arbitrary.

To construct the calibration curve from the data (s_{ej}, i_{ej}), the least-squares adjustment method can be used. This is explained in section 5.6. Examples of calibration curves are given in sections 5.8.2 and 5.8.4.

3.11.4 Verification and adjustment

The owner or user of a measuring instrument may only be interested in its *verification* or *confirmation*. These terms (not defined in the VIM) usually refer to the process by which it is checked whether the absolute value of the difference between the indications G and the corresponding values of standards S that cover the instrument's range of measurement is no greater than a certain given value ε, known as either the *accuracy* or *maximum permissible error* of the instrument. Alternatively, it might be desired to check that the value of the ratio S/G is not greater than $1 + \phi$ nor smaller than $1 - \phi$, where ϕ is the allowable *nonlinearity* (figure 3.32). If the instrument satisfies these conditions, a rectangular pdf centred at zero and width 2ε will adequately describe the state of knowledge about the calibration correction C, while a rectangular pdf centred at one and width 2ϕ will adequately describe the state of knowledge about the calibration correction factor F. Then, from the models

$$Q = G + C$$

or

$$Q = FG$$

with $c_e = 0$ or $f_e = 1$, the user simply obtains

$$q_e = g_e$$

and

$$u_q^2 = u_g^2 + u_c^2 = u_g^2 + \frac{\varepsilon^2}{3}$$

or

$$u_q^2 = (f_e u_g)^2 + (g_e u_f)^2 = u_g^2 + \frac{(g_e \phi)^2}{3}.$$

Thus, as defined here, the process of verification is a simpler form of calibration.

Calibration and verification (or confirmation) should not be confounded with *adjustment*. In fact, before proceeding to calibrate an instrument, it should be adjusted according to the manufacturer's specifications. The VIM's definition of adjustment is

> operation of bringing a measuring instrument into a state of performance suitable for its use.

Sometimes the adjustment operations are similar to those performed in calibration. For example, consider an instrument that has two adjustment controls to modify independently the indications corresponding to two given values of measurands to which the instrument is applied. Suppose that a pair of standards S_a and S_b of known values s_a and s_b are used. We apply standard S_a and use the first control to make the indication pass from g_{a1} to $g_{a2} = s_a$. Next, we apply standard S_b and use the second control to make the indication pass from g_{b2} to $g_{b3} = s_b$ (g_{b1} is the indication that would be obtained for standard S_b with no adjustment, while $g_{a3} = g_{a2}$ is the indication that would be obtained for standard S_a after both adjustments).

The procedure is illustrated in figure 3.33, where A, B and C are the calibration curves that would be obtained at the three control settings. But

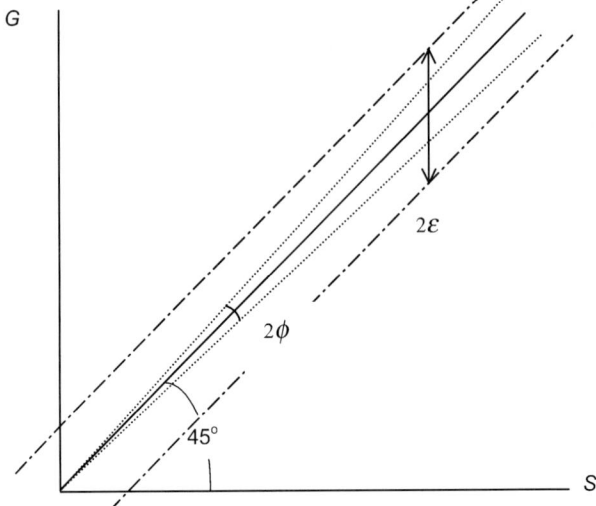

Figure 3.32. Maximum permissible error ε and allowable nonlinearity ϕ.

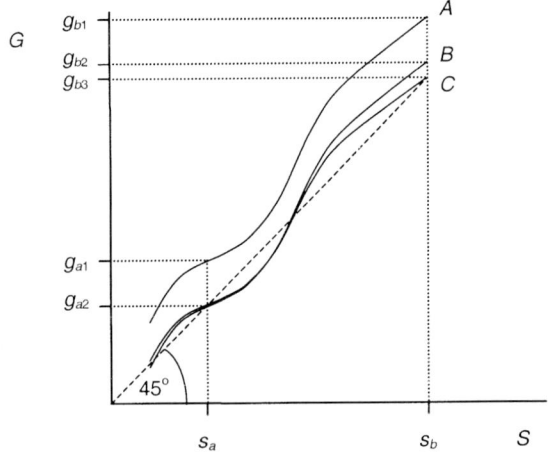

Figure 3.33. An adjustment process.

note that the adjustment gives no information whatsoever for constructing these curves. All that was done was to set the instrument error equal to zero at the two adjustment points.

Of course, not all instruments have both types of control. Sometimes, they have only a *zeroing* control to set the instrument's zero (i.e. to set $g = 0$ when $s = 0$). If the second control is available, it modifies the instrument's *gain*, that is the slope of the input–output curve. In weighing instruments, the process of setting $g = 0$ when S has any value is called *taring* (for example, the fourth step in the method for measuring the volume V_f° of the flask in section 3.10 is a taring procedure).

Example. Consider wrist watches. These instruments are never calibrated, but most of us do *adjust* them from time to time using radial or telephone information as the (not necessarily traceable) standard.

Chapter 4

Evaluating the expanded uncertainty

Even though the pdf $f(q)$ encodes the state of knowledge about a measurand Q, the metrologist will not usually communicate the pdf itself but two of its parameters: the expectation value and the standard deviation. The former is taken as the best estimate or most representative value of the quantity, the latter is taken as the standard uncertainty and represents the typical spread of the pdf about the estimated value. Both parameters are intrinsic properties of $f(q)$, and thus depend only on the information given about Q. If properly determined, the standard uncertainty serves to characterize the quality of the measurement.

In practice, however, users of the measurement result will often not be interested in the standard uncertainty *per se*. For decision purposes, for example in conformity assessment, they require an interval that can be viewed as containing the value of the measurand with a relatively high probability of being right. The half-width of this interval is known as the *expanded uncertainty*. Its evaluation is the subject of the present chapter.

More specifically, in section 4.1 the important concepts of coverage factor and coverage probability are first defined rigorously. These are then related to each other for the case of some of the pdfs that appear commonly in practice. Next, section 4.2 defines the more general concept of coverage uncertainty, to which the expanded uncertainty corresponds when the pdf is symmetric.

Since the expanded uncertainty is mainly used in conformity assessment, one needs to understand that the probabilities of rejecting an acceptable item and of accepting a non-conforming item are always present. These are defined in section 4.3 as the producer's and consumer's risks, respectively. In general, since the latter is more detrimental than the former, a procedure known as guardbanding is applied. This involves taking the acceptance interval as that obtained by subtracting twice the expanded uncertainty from the specification interval that applies for the tested product. In such a way, as shown in section 4.4, the consumer's risk is reduced.

We turn next to the classical procedures for determining the expanded uncertainty. This is done in section 4.5 for a directly measured quantity and in section 4.6 for a quantity that is evaluated from a measurement model. It is shown that these procedures appear to be doubtful, especially when the model involves quantities that are not evaluated statistically. The chapter ends by asserting that, contrary to the prevalent view, the expanded uncertainty may be rather superfluous in practice.

4.1 Coverage factor and coverage probability

For a measurand Q, the expanded uncertainty was defined in section 3.2.3 as the product $U_{pq} = k_p u_q$ where u_q is the standard uncertainty and k_p is a number known as the *coverage factor*. The latter is not just any constant: its subscript is meant to emphasize that it is associated with a *coverage probability* defined as

$$p = \int_{q_e - U_{pq}}^{q_e + U_{pq}} f(q) \, dq.$$

The coverage probability should by no means be interpreted as the expected proportion of times that the value of Q will be found to lie within the interval $(q_e - U_{pq}, q_e + U_{pq})$ each time Q is measured. The quantity may, in fact, be measured repeatedly but the information thus gathered is used to find the measurement estimate and the standard uncertainty. Rather, p should be interpreted as the degree of belief ascribed to the (true) value of Q being inside the interval.

The quantitative relation between p and k_p depends, of course, on the pdf $f(q)$ that is obtained from the measurement information. Let us consider some examples.

4.1.1 The normal pdf

The expression for a normal pdf centred about the value q_e and variance u_q^2 is

$$N(q) = \frac{1}{\sqrt{2\pi} u_q} \exp\left[-\frac{1}{2} \frac{(q - q_e)^2}{u_q^2}\right].$$

Now recall equation (3.17)

$$\int_{q_a}^{q_b} N(q) \, dq = \Phi(x_b) - \Phi(x_a)$$

where $x_a = (q_a - q_e)/u_q$, $x_b = (q_b - q_e)/u_q$ and $\Phi(x)$ is the standard normal cumulative function. If we write $x^* = (q^* - q_e)/u_q$, then

$$\int_{-q^*}^{q^*} N(q) \, dq = 2\Phi(x^*) - 1. \tag{4.1}$$

Table 4.1. Coverage factors for the normal pdf for various coverage probabilities.

$p/\%$	k_p
50	0.675
66.67	0.968
68.27	1
75	1.15
85	1.439
90	1.645
95	1.96
95.45	2
98	2.327
99	2.576
99.73	3

Let now $q^* = q_e + U_{pq}$. Then $x^* = k_p$ and we have

$$p = 2\Phi(k_p) - 1.$$

This relation is tabulated for convenience in table 4.1 for some selected values of p and k_p.

Example. To find $k_{0.75}$ for the normal pdf we first calculate $\Phi(k_{0.75}) = (0.75 + 1)/2 = 0.875$. Then $k_{0.75} = 1.15$ is obtained from table A1.

Example. According to the calibration certificate of a 10 Ω standard resistor, its conventional true resistance is $10.000\,742\ \Omega \pm 129\ \mu\Omega$, where the expanded uncertainty 129 μΩ defines a coverage probability of 99 %. From table 4.1 the coverage factor is 2.576 and the standard uncertainty is $(129/2.576 = 50)$ μΩ.

4.1.2 Trapezoidal, rectangular and triangular pdfs

The symmetric trapezoidal pdf is depicted in figure 3.6. For this pdf, the following expression is easily derived (see figure 4.1):

$$\frac{U_{pq}}{e} = \begin{cases} p(1+\phi)/2 & \text{for } p < p_{\lim} \\ 1 - \sqrt{(1-p)(1-\phi^2)} & \text{for } p > p_{\lim} \end{cases}$$

where

$$p_{\lim} = \frac{2\phi}{1+\phi}.$$

Coverage factor and coverage probability

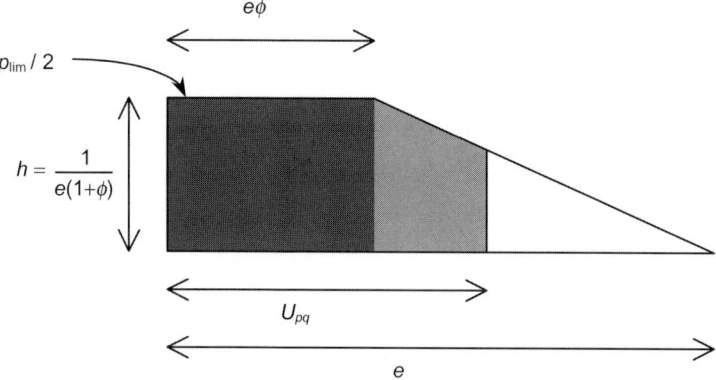

Figure 4.1. The symmetric trapezoidal pdf.

Hence, recalling from equation (3.18), that

$$\frac{u_q}{e} = \sqrt{\frac{1+\phi^2}{6}}$$

we get, for $p < p_{\lim}$,

$$k_p = \frac{\sqrt{6}(1+\phi)}{2\sqrt{1+\phi^2}} p \qquad (4.2)$$

and, for $p > p_{\lim}$,

$$k_p = \frac{1-\sqrt{(1-p)(1-\phi^2)}}{\sqrt{(1+\phi^2)/6}}. \qquad (4.3)$$

For the triangular pdf, $p_{\lim} = 0$ and equation (4.2) applies with $\phi = 0$:

$$k_p = \sqrt{6}\left(1 - \sqrt{1-p}\right).$$

For the rectangular pdf, $p_{\lim} = 1$ and equation (4.3) applies with $\phi = 1$:

$$k_p = \sqrt{3}p.$$

4.1.3 Student's-t pdf

The pdf

$$S_\nu(t) = K_\nu \left(1 + \frac{t^2}{\nu}\right)^{-(\nu+1)/2}$$

is called Student's-t with ν degrees of freedom. The expression for the normalization constant K_ν is not important for our purposes, but it is nevertheless

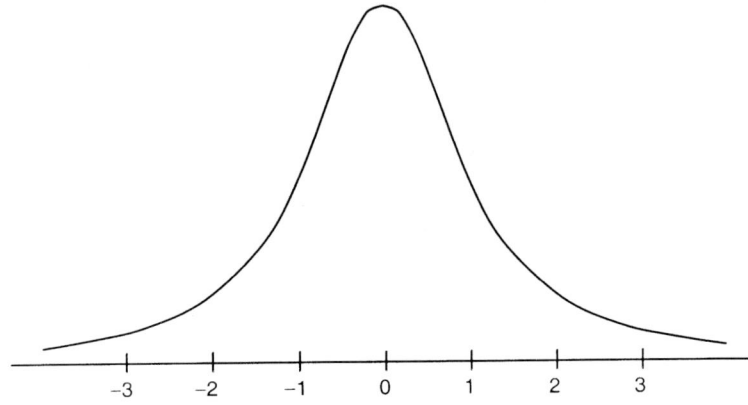

Figure 4.2. The t-Student pdf with one degree of freedom.

given here for completeness:

$$K_\nu = \frac{\Gamma[(\nu+1)/2]}{\Gamma(\nu/2)\sqrt{\pi\nu}}$$

where Γ is the gamma function

$$\Gamma(y) = \int_0^\infty x^{y-1} e^{-x}\, dx$$

defined for $y > 0$.

The t-density is symmetric and its shape approaches that of the normal as ν increases. Its expectation is zero and its variance is

$$r^2 = \frac{\nu}{\nu - 2} \qquad (4.4)$$

defined only for $\nu > 2$ (for ν equal to one or two neither the variance nor the standard uncertainty can be defined). For one degree of freedom, the t-density is

$$S_1(t) = \frac{1}{\pi(1+t^2)}.$$

This density is depicted in figure 4.2.

Values $t_{p,\nu}$ satisfying

$$\int_{-t_{p,\nu}}^{t_{p,\nu}} S_\nu(t)\, dt = p \qquad (4.5)$$

are given in table A2 in the appendix. These values will here be called t-quantiles. Therefore, if the pdf of

$$T = \frac{Q - q_e}{u_q}$$

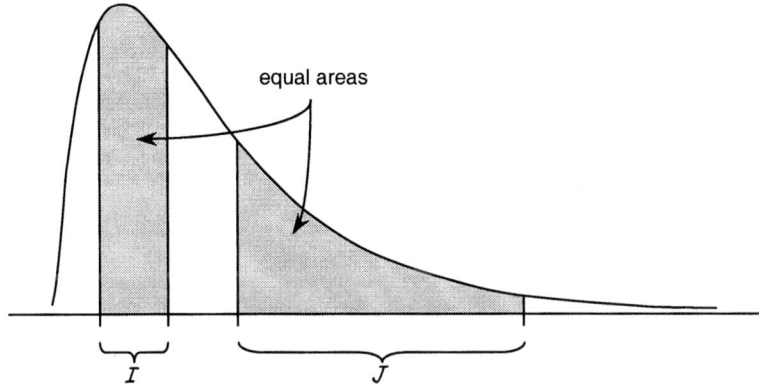

Figure 4.3. The coverage probabilities for the intervals I and J are equal.

is t-Student with $\nu \geq 2$, $t_{p,\nu}$ becomes identical to the coverage factor k_p. The usefulness of the t-density will be explained in section 4.5.

4.2 Coverage intervals and coverage uncertainties

As previously defined, the expanded uncertainty cannot be calculated if the standard uncertainty does not exist. This occurs, for example, in the case of Student's-t pdf with one degree of freedom depicted in figure 4.2. A more general concept is that of *coverage uncertainty*, whose definition (not given in the *Guide*) is based on that of *coverage intervals*. These are defined as the intervals I_p over which the integral of the pdf is equal to p. In symbols

$$p = \int_{I_p} f(q)\,dq.$$

It should be clear that this equation does not fix I_p. For example, the intervals I and J in figure 4.3 define equal coverage probabilities, in the normal density, the equation $p = \Phi(x_b) - \Phi(x_a)$ does not define a unique pair of values x_a and x_b, in a rectangular density of width δ, any coverage interval of length $p\delta$ within the domain of the density has an associated coverage probability equal to p.

To fix the coverage interval, it seems reasonable to impose the conditions that its limits should encompass the best estimate, and that equal coverage probabilities should correspond to both sides of the estimate. Evidently, this means that for symmetric pdfs the coverage interval will also be symmetric about the best estimate. Otherwise, both its limits must be calculated independently.

Once I_p has been obtained, the *right* and *left* coverage uncertainties may be defined as the distances from the upper and lower limits of the interval, respectively, to the best estimate.

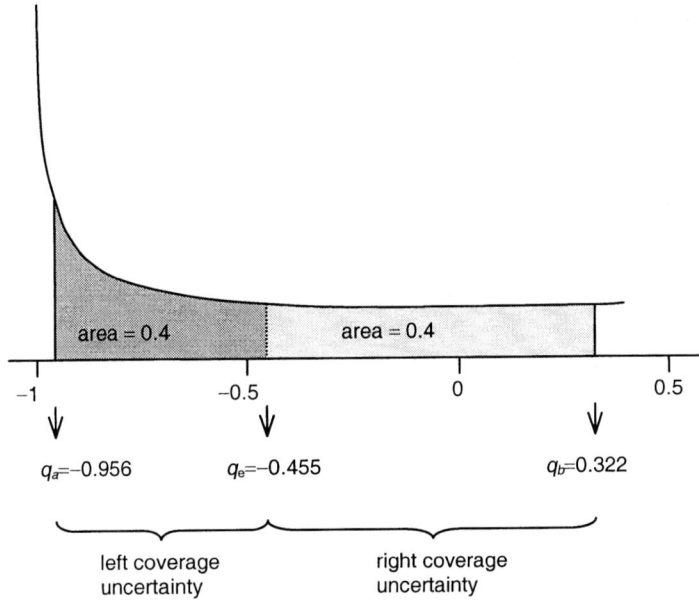

Figure 4.4. Coverage uncertainties for an asymmetric pdf.

Example. Consider the 'distorted U' pdf, equation (3.33)

$$f(q) = \frac{1}{c\sqrt{(q-a)[2b-(q-a)]}}$$

for $a < q < a+b(1-\cos c)$, where a, b and c are constants, $b > 0$ and $c \leq \pi$. It is easy to show that for a coverage probability p, the limits of the interval that satisfy the stated conditions are

$$q_b = a+b-b\cos\left[\cos^{-1}\left(\frac{a+b-q_e}{b}\right) + \frac{pc}{2}\right]$$

and

$$q_a = a+b-b\cos\left[\cos^{-1}\left(\frac{a+b-q_e}{b}\right) - \frac{pc}{2}\right].$$

where q_e is evaluated using (3.30). These limits are shown in figure 4.4 for $a = -1, b = 1, c = 2$ and $p = 80\%$.

Since the coverage interval can always be calculated, the coverage uncertainties exist irrespective of whether or not the standard uncertainty is defined. If the pdf is symmetric, both coverage uncertainties are equal. If the pdf is symmetric and its variance exists, the coverage factor may be redefined as the ratio of the coverage uncertainty to the standard uncertainty.

Example. Consider the Student's-t pdf with one degree of freedom. The limits of the symmetric interval I_p turn out to be $\pm\tan(\pi p/2)$. Since the expectation of this pdf is zero, both coverage uncertainties are equal to $\tan(\pi p/2)$. However, the coverage factor is undefined.

4.3 The consumer's and producer's risks

From the examples in section 4.1, it may be seen that for $I_p = (q_e - u_q, q_e + u_q)$ the coverage probability is usually not very large. The reader may verify that this is 68.3% for the normal density, 57.7% for the rectangular density and 65.9% for the t-density with 10 degrees of freedom. This is precisely the reason for expressing the expanded uncertainty with a coverage factor larger than one: to obtain an interval about the estimate q_e where the value of the measurand is expected to lie with a 'reasonable degree of certainty'. Having done this, sometimes the measurement outcome will be reported as $q_e \pm U_{pq}$, although such notation may be misleading, see section 4.7.

Now, what is meant by a coverage probability being 'reasonably large'? Return to our introductory example in chapter 2. Suppose that, without doing any measurements or calculations, a quality inspector estimates that the length of a spacer rod is less than 1 km, with $p = 1$. Obviously the inspector is right, but there is also no question that such 'result' is not reasonable. Then again, suppose that the inspector is more thorough: after measurement, he/she obtains a best estimate $q_e = 9.949$ mm and concludes that an expanded uncertainty equal to 0.025 mm gives a 90% coverage probability, while a 0.029 mm expanded uncertainty gives a 95% coverage probability. Which coverage interval is more 'reasonable'?

The problem the inspector confronts is typical of many manufacturing operations. In most of them it has to be checked that product characteristics conform to their specifications. As discussed in chapter 2, the specifications are usually cast in the following way: for the product to be accepted, the true value q_t of a given characteristic must be within a certain specification interval S. Suppose this interval is symmetric, of the form $S = (\mu - \varepsilon, \mu + \varepsilon)$ where μ is the desired or nominal value and 2ε is the tolerance. This means that only those products that satisfy the condition $\mu - \varepsilon \leq q_t \leq \mu + \varepsilon$ should be accepted. For example, the specification for the spacer might be $\mu = 10$ mm and $\varepsilon = 0.1$ mm. Thus, the spacer would pass the inspection if q_t was known to be greater than 9.9 mm and less than 10.1 mm. Now the problem is that, whereas the parameters μ and ε are given, and thus known, the value q_t is only *assumed* to be equal to the estimated (measured) value q_e. As a consequence, wrong acceptance or rejection decisions may unknowingly be made.

For example, let us suppose that, after taking into account all corrections for systematic effects, the value $q_e = 9.949$ mm is obtained for the spacer's length. This value falls within the specification interval S. However, if (unknowingly) the rod's length was actually $q_t = 9.897$ mm, say, the product would be *incorrectly accepted*. The probability that such an event occurs is known as the 'consumer's risk', because it relates to the quality of the product as perceived by its users, be they the final customers or the workers down the production line. If they receive too short a spacer, it will fit loosely. If the spacer is too long, it will not fit at all.

However, the value q_e might be found to lay outside the interval S while q_t does actually (but unknowingly) satisfy the specification criterion. This combination would lead to a *false rejection* decision. The probability of incorrectly rejecting an item is called the 'producer's risk', because its consequences (unnecessary costs) fall mainly on the manufacturer.

In practice, the consumer's risk is usually deemed to be more harmful than the producer's risk. This is because whereas the costs of rejecting an acceptable item are assumed only by the manufacturer, the decision to accept a faulty product will have deleterious consequences not only on its users, but also on the perceived reliability of the producer. In most cases, the purpose of calculating and using the expanded uncertainty is precisely to reduce the consumer's risk. The general idea—already introduced in section 2.4—is as follows.

4.4 The acceptance interval: guardbanding

Let μ again be the nominal value of the inspected characteristic for the population of products and ε the half-width of the specification interval S. Now define the acceptance interval as $A_p = (\mu - \kappa_p \varepsilon, \mu + \kappa_p \varepsilon)$ where κ_p is a positive number sometimes known as the 'guardband factor'. If κ_p is chosen to be less than one, the acceptance interval becomes *smaller* than the specification interval, and the former replaces the latter in the acceptance criterion. This is called 'guardbanding', see figure 4.5.

The factor κ_p may be established conveniently from the expanded uncertainty U_{pq}. Suppose that the best estimate complies with $\mu - \varepsilon + U_{pq} \leq q_e \leq \mu + \varepsilon - U_{pq}$ and that the coverage probability is large, such that there is a small chance that $q_t \leq q_e - U_{pq}$ or $q_t \geq q_e + U_{pq}$. Consequently, there will be an even smaller chance that $q_t < \mu - \varepsilon$ or $q_t > \mu + \varepsilon$. Therefore, it seems appropriate to take

$$\kappa_p \varepsilon = \varepsilon - U_{pq}$$

or

$$\kappa_p = 1 - \frac{k_p u_q}{\varepsilon}.$$

This criterion is recommended in [23]. However, it should be noted that adopting the interval A_p as the acceptance criterion does not *eliminate* the consumer's risk.

Example. For a spacer rod let $\mu = 10$ mm and $\varepsilon = 0.1$ mm, such that $S = (9.9, 10.1)$ mm, and suppose that $q_e = 9.949$ mm with $U_{0.9q} = 0.025$ mm. The guardband factor becomes $\kappa_{0.9} = 0.75$,

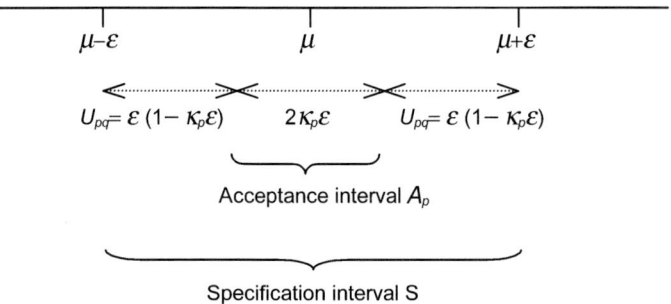

Figure 4.5. The specification and acceptance intervals.

therefore $A_{0.9} = (9.925, 10.075)$ mm and the rod will be accepted. But since there is a 5% chance that $q_t < (9.949 - 0.025)$ mm $= 9.924$ mm, there is also a non-zero chance that $q_t < 9.9$ mm. The latter (false acceptance) probability would be reduced by taking $U_{0.95q} = 0.029$ mm, but then the width of the acceptance interval would decrease correspondingly.

It is thus seen that the 'reasonableness' of the expanded uncertainty depends on a tradeoff between a very small but 'sure' acceptance interval against the sizes of A_p and S being almost equal but with a larger consumer's risk. The quantitative evaluation of the consumer's and producer's risks will be taken up in section 6.7. We turn now to the evaluation of coverage probabilities.

4.5 Type A analysis revisited

The TA procedure, as described in section 3.8, applies to a quantity Q that is measured n times independently, under repeatability conditions. The result is the *random sample* $\theta = (q_1, \ldots, q_n)$. Under the framework of classical statistics, the quantity Q is regarded as a fixed non-varying quantity having an unknown value μ. The observations are described by the *random sequence* $\Theta = (Q_1, \ldots, Q_n)$, where the Q_is are independent random variables having the same underlying *frequency* distribution $f(q)$ with expectation μ and unknown variance σ^2. Two important functions of the random sequence, the *sample mean*

$$\overline{Q} = \frac{1}{n} \sum_{i=1}^{n} Q_i$$

and the *sample variance*

$$S^2 = \frac{1}{n-1} \sum_{i=1}^{n} (Q_i - \overline{Q})^2$$

assume the values

$$\bar{q} = \frac{1}{n} \sum_{i=1}^{n} q_i$$

and

$$s^2 = \frac{1}{n-1} \sum_{i=1}^{n} (q_i - \bar{q})^2$$

respectively. Since $E(\overline{Q}) = \mu$, $E(S)^2 = \sigma^2$ and $V(\overline{Q}) = \sigma^2/n$, we take $q_e = \bar{q}$ and $u_q = s/\sqrt{n}$.

The foregoing results are valid whatever the distribution $f(q)$. Assume now that this distribution is normal with parameters μ and σ^2. It can be shown that, in this case, the distribution of $\overline{Q}$ is normal as well, with the same expectation and with variance σ^2/n. This means that we can write

$$f(\bar{q}) = \frac{1}{\sigma}\sqrt{\frac{n}{2\pi}} \exp\left[-\frac{n}{2\sigma^2}(\bar{q} - \mu)^2\right]$$

where $\bar{q}$ should be interpreted here as a *variable*, not as the one value that is obtained after the experiment has been completed. Thus, from (4.1), the probability of obtaining a sample mean $\bar{q}$ within the interval $(\mu - k_p\sigma/\sqrt{n}, \mu + k_p\sigma/\sqrt{n})$ is equal to $p = 2\Phi(k_p) - 1$.

However, this probability is only of theoretical interest, because μ and σ^2 are not known. Nevertheless, for the sake of the following discussion, assume that the variance σ^2 is, in fact, given. Since the inequalities

$$\mu - k_p\sigma/\sqrt{n} < \bar{q} < \mu + k_p\sigma/\sqrt{n}$$

are equivalent to

$$\bar{q} - k_p\sigma/\sqrt{n} < \mu < \bar{q} + k_p\sigma/\sqrt{n}$$

one might think of ascribing the probability p to the event of finding the unknown expectation μ within the interval $I_p = (\bar{q} - k_p\sigma/\sqrt{n}, \bar{q} + k_p\sigma/\sqrt{n})$. This would be quite incorrect, however, because μ is a parameter, not a random variable. Given the particular sample mean $\bar{q}$ that is calculated from the repeated measurements, the unknown value μ will either be within I_p or not. Rather, p should be interpreted as the *proportion* of intervals that would be expected to contain the value μ were many sample means to be computed from a number of samples of the same size. Therefore, strictly, p is *not* a coverage probability. It is called the *confidence level* associated with the *confidence interval* I_p for the parameter μ.

Example. Let Q be the life length of a certain piece of equipment. One hundred pieces are tested, yielding an average $\bar{q} = 501.2$ h. Assume that the pdf of Q is normal with unknown expectation μ and known standard deviation $\sigma = 4$ h. To obtain the confidence interval for μ

with confidence level $p = 95\%$ we find $k_{0.95} = 1.96$ from table 4.1. The interval's limits (in hours) are then

$$501.2 - 1.96\frac{4}{\sqrt{100}} = 500.4$$

and

$$501.2 + 1.96\frac{4}{\sqrt{100}} = 502.0.$$

Now in practice the variance σ^2 will not normally be a datum. We then use again $E(S^2) = \sigma^2$ and the fact that, if $f(q)$ is normal, the frequency distribution of the statistic

$$T = \frac{\sqrt{n}}{S}(\overline{Q} - \mu)$$

is a Student's-t with $\nu = n - 1$ degrees of freedom. Therefore, from (4.5), the probability of obtaining a sample mean $\overline{q}$ within the interval $(\mu - k_p s/\sqrt{n}, \mu + k_p s/\sqrt{n})$ is evaluated using $k_p = t_{p,n-1}$ (or, since Student's distribution approaches the normal shape as n increases, table 4.1 might be used for n greater than about 50, see the bottom lines of table A2). Following the same earlier reasoning, and recalling that u_q is taken as $s/\sqrt{n}$, we conclude that the confidence interval for μ with unknown variance σ^2 becomes $(\overline{q} - t_{p,n-1}u_q, \overline{q} + t_{p,n-1}u_q)$.

Example. Ten measurements of a resistance Q are taken. The results yield an average $\overline{q} = 10.48\ \Omega$ and a precision $s = 1.36\ \Omega$. We then have $u_q = 1.36/\sqrt{10} = 0.43\ \Omega$. For a confidence level $p = 90\%$ we get from table A2 $t_{0.90,9} = 1.83$. The corresponding confidence interval is bound by the values (in Ω)

$$10.48 - 1.83 \times 0.43 = 9.69$$

and

$$10.48 + 1.83 \times 0.43 = 11.27.$$

Suppose now that the values $\overline{q}$ and s are calculated from $n = 200$ measurements. Then, the standard uncertainty decreases to $u_q = 1.36/\sqrt{200} = 0.10\ \Omega$. The confidence interval is also much reduced. This is now calculated using not the t-density but the normal pdf, i.e. table 4.1, from which $k_{0.90} = 1.645$. The new bounds (in Ω) of the confidence interval are then

$$10.48 - 1.645 \times 0.10 = 10.32$$

and

$$10.48 + 1.645 \times 0.10 = 10.64.$$

This way of calculating the expanded uncertainty depends on $f(q)$ being normal. If there are reasons to believe that the values q_i are sampled from another frequency distribution, other statistical procedures should be used to compute the coverage factor. This will not be needed for samples of large size, however, because the *central limit theorem* (CLT, see later) allows one to assume normality for the sample mean $\overline{Q}$ whatever the distribution $f(q)$. Of course, in metrology such large samples are seldom available.

4.6 Quantities evaluated from a measurement model

Normally one should not have much use for an expanded uncertainty calculated as before, because it applies only to a quantity that will, in general, be an input to the model of the measurand Q. While in such cases we do have a simplified way of determining the standard uncertainty (see section 3.4), the evaluation of the coverage factor requires that $f(q)$ be known. An approximate expression for this distribution follows from the CLT, as will now be explained.

Assume the measurand is evaluated according to a measurement model $Q = \mathcal{M}(Z_1, \ldots, Z_N)$ involving N input quantities. Assume also that the model is not too strongly nonlinear, such that the linearized quantity

$$Q_L = \mathcal{M}(\mu_1, \ldots, \mu_N) + \sum_{i=1}^{N} c_i (Z_i - \mu_i)$$

deviates little from Q in the neighbourhood of $\mathcal{M}(\mu_1, \ldots, \mu_N)$, where the μ_is are the expectations of the input quantities and the c_is are the sensitivity coefficients $\partial \mathcal{M} / \partial Z_i$ taken at the expectations. Assume further that the input quantities are all independent and that N is large. The CLT asserts that the frequency distribution of Q_L (and consequently of Q) will be approximately normal, centred at $\mathcal{M}(\mu_1, \ldots, \mu_N)$ and variance

$$\sigma_q^2 = \sum_{i=1}^{N} c_i^2 \sigma_i^2$$

where the σ_i^2 are the variances of the input quantities.

As discussed in section G.2 of the *Guide*, the convergence to normality will be faster: (i) the larger the number N of input variables is, (ii) the closer the products $c_i \sigma_i$ are to each other and (iii) the closer the distributions of the Z_i's are to being normal. For example, in Clause G.2.2 of the *Guide* it is shown that the distribution corresponding to a sum of as few as three quantities described by rectangular distributions of equal width is quite well approximated by the normal.

Assume then that the model $\mathcal{M}(Z_1, \ldots, Z_N)$ is a not too strongly nonlinear function of the input quantities and that these are such that the requirements for the application of the CLT are satisfied. Suppose further that the information pertaining to all input quantities is processed according to the TA procedure. This implies that the unknown expectations μ_i will be estimated as the sample means $z_{ei} = \bar{z}_i$. The *Guide* recommends evaluating the best estimate of Q as $q_e = \mathcal{M}(z_{e1}, \ldots, z_{eN})$.

The procedure also implies that the unknown variances σ_i^2 will be estimated as the sample variances s_i^2. However, the estimate of σ_q^2 is not taken as the sum of the products $c_i^2 s_i^2$ (where the c_is are now taken at the estimates z_{ei}). Rather, because of the LPU, Clause G.4.1 of the *Guide* recommends expressing the standard uncertainty associated with q_e as the square root of

$$u_q^2 = \sum_{i=1}^{N} c_i^2 u_{zi}^2 \tag{4.6}$$

with $u_{zi}^2 = s_i^2/\sqrt{n_i}$, n_i being the number of measurements of the quantity Z_i. In the opinion of this author, (4.6) stands on feeble theoretical grounds, because it appears to contradict the CLT. Note also that the n_is are not all necessarily equal to each other, a fact that puts (4.6) further into question. Nevertheless, it appears to work well most of the time.

The recommended procedure for calculating the corresponding coverage factor is also based on the assumption that the frequency distribution associated with Q is normal. One should then take k_p as equal to a t-factor calculated from Student's distribution with an 'effective' number of degrees of freedom computed according to the Welch–Satterthwaite (WS) formula

$$\nu_{\text{eff}} = u_q^4 \left[\sum_{i=1}^{N} \frac{(c_i u_{zi})^4}{\nu_i} \right]^{-1} \tag{4.7}$$

where $\nu_i = n_i - 1$. However, the *Guide* does not explain why the use of the WS formula should be appropriate in this context. In fact, it has been shown [2] that this formula may produce sometimes unreasonable results. Nevertheless, other investigators advocate its use [18].

Up to this point the discussion has been limited to the case when all input uncertainties are derived from frequential information. However, in practice the state of knowledge about some or even all of the input quantities is often not of this kind. When this occurs, the *Guide* recommends resorting to the so-called 'type B evaluation', by which the standard uncertainties are derived as the standard deviations of the corresponding probability densities, as explained in section 3.2. In this case it is proposed that we still use equations (4.6) and (4.7), assigning a number of degrees of freedom to each type B evaluated uncertainty. These degrees of freedom are intended to quantify 'how uncertain' one is about the non-frequentially determined standard uncertainties. The rule is

$$\nu_i = \frac{1}{2(1 - R_i)^2} \tag{4.8}$$

where R_i is defined as the 'reliability' of the uncertainty u_{zi}. In the *Guide*, the factor $1 - R_i$ is termed 'the relative uncertainty of u_{zi}' and is written as $\Delta u_{zi}/u_{zi}$. Once all degrees of freedom have been computed, either as $n_i - 1$ or from (4.8), the WS is then applied to calculate ν_{eff}.

Example. If it is judged that u_{zi} is reliable to 75 % (equivalent to 25 % relative uncertainty of the uncertainty) the number of degrees of freedom ν_i is eight. If instead one is quite 'sure' that u_{zi} does in fact represent a 'reliable' standard uncertainty, then $R_i \to 1$ and $\nu_i \to \infty$.

Example. Let $Q = X + Y$ where X is described by a rectangular pdf of half-width e_x and Y is repeatedly measured n times yielding a precision s. If it is assumed that the half-width e_x is 100 % reliable, the effective degrees of freedom are

$$\nu_{\text{eff}} = \frac{n^2(n-1)}{s^4}\left(\frac{e_x^2}{3} + \frac{s^2}{n}\right)^2.$$

At least to the present author this procedure, which might be called 'frequentialization' for lack of a better term, does not appear very convincing. In addition to the inconsistencies mentioned earlier we note the following objections:

- The *Guide* is silent about the validity of (4.6) when the requirements for the application of the CLT are not satisfied, in particular when the measurement model does not involve a large number of input quantities or if their contributions to the uncertainty of the output quantity is much dissimilar.
- The procedure does not yield a pdf for the measurand and therefore the interpretation of the standard uncertainty as a standard deviation is doubtful.
- The use of the t-distribution in this context will not render a coverage probability but a confidence level.
- The procedure recommended in the *Guide* does not allow one to take into account any prior information about the output quantity that may be available (for example, if the measurand refers to a calibrand its previous calibration certificate may be used; or if it stands for a measurable characteristic of a certain product prior knowledge may be taken from the statistical records of the production line).

One further difficulty concerns what to do if the input quantities are not independent. Unfortunately, in this respect the *Guide* runs out of easy shortcuts. It only reminds us, in Clause F.1.2.4, about the obvious fact that, should correlations arise because one or more other common quantities enter into the model relations for the quantities Z_is, these correlations can be eliminated by remodelling Q in terms of independent input quantities only (recall that this approach was taken in the example in section 3.10, see also [30]).

Among all these objections, however, the use of (4.8) is probably the most questionable. This equation arises from forcing a pdf to be interpreted as a frequency distribution. Since the latter is a limiting concept that cannot be exactly known, an uncertainty about its 'true' standard deviation exists. However, the very expression 'uncertainty of the uncertainty' implies a contradiction in terms. Indeed, to speak of 'estimating' an uncertainty implies that there exists a 'true' uncertainty, which is obviously absurd. But even if one is willing to consider these as theoretical subtleties, there remains the operational problem of establishing how 'uncertain' an uncertainty is. And if this is accepted, should there not also be an 'uncertainty in the uncertainty of the uncertainty'? By which rules should each uncertainty in this (infinite) chain be obtained? How should these uncertainties be compounded?

Most of the problems and inconsistencies that appear using the approach in the *Guide* seem to arise because sampling theory does not distinguish between a frequency distribution and a pdf. The former is regarded as a mathematical entity that, in principle, could be completely known if unlimited repetitions of the random experiment were to be carried out. The fact that this is not done leads to an uncertainty about the 'actual' shape of the distribution or, equivalently, to an uncertainty about its 'true' standard deviation.

In contrast, suppose that the state of knowledge about each input quantity was encoded through an appropriate pdf. The rules of probability calculus could then be applied to obtain the pdf corresponding to the output quantity. The output standard uncertainty would be identified with the standard deviation of the resulting pdf and it would relate only to the value of the *measurand*, not to the pdf, which would be regarded as completely *known*. Therefore, the question is: Are there objective criteria by which input pdfs may be obtained? Would these rules apply as well to quantities for which statistical information is available? Under the framework of *Bayesian statistics*, the answer to both questions is 'yes'. These matters will be addressed in chapter 6.

4.7 Standard *versus* expanded uncertainty

In this chapter, the usefulness of calculating expanded uncertainties and coverage probabilities was seen to follow from the need to define 'reasonable' acceptance intervals. Since this task may become considerably difficult, it is important to reflect on whether it should always be carried out. In the opinion of this author, calculating (and reporting) expanded uncertainties makes sense only in applications where measurement results are to be compared against specifications.

Take, for example, material properties. Suppose the measurand is the modulus of elasticity E of some given metal specimen. A typical result might be $E_e = 100$ GPa with a 10% relative standard uncertainty, equivalent to $u_E = 10$ GPa. The research team that made this evaluation might make further processing of the available information and apply the CLT in conjunction with the WS formula to conclude that the degrees of freedom are sufficiently large to warrant the use of the normal distribution for calculating the coverage factor.

They might make use of table 4.1 to calculate an expanded uncertainty $U_E = 20$ GPa for a coverage probability of about 95 %. The measurement result would then be reported as $E_e = (100 \pm 20)$ GPa.

This way of reporting a measurement result, while better than giving just a point value, is highly undesirable, because it tends to communicate the idea that the interval (80 GPa, 120 GPa) comprises *all* possible values of the modulus of elasticity. A better statement might be 'the estimated value of the modulus of elasticity is 100 GPa with an expanded uncertainty of 20 MPa, where the expanded uncertainty was calculated using a coverage factor equal to 2 and defines a coverage interval giving a coverage probability of about 95 %'.

Now reflect for the moment on who might want to use this information. Normally, material properties such as the modulus of elasticity are not of interest in themselves, but are needed to evaluate the behaviour of the material when subject to a certain load. For example, the modulus of elasticity might be needed to calculate the deflection of a beam made of the specimen's metal under some force. The deflection would then become 'the' measurand, to which the modulus of elasticity is just one of the input quantities. The uncertainty in the beam's deflection should then be calculated using an equation such as (3.22), for which the *standard*, not the expanded uncertainty, is required. The person who evaluates the deflection would need to divide the reported expanded uncertainty in knowing E by the stated coverage factor. Therefore, all the trouble taken to evaluate the expanded uncertainty would have been for nothing.

Another example is found in calibration certificates. As explained in section 3.11, these certificates provide information on corrections or correction factors applicable to the calibrand. These quantities are almost *never* of interest in themselves, but are needed as input quantities to obtain corrected measurement results. Similarly, it would be of little use to calculate the expanded uncertainties associated with the estimated coefficients of a least-squares adjusted curve, since they are normally needed to predict the value of one coordinate when the other is given. In the computation of the resulting uncertainty, only the standard uncertainties of the coefficients are needed (see the examples in sections 5.8.2 and 5.8.4).

One then wonders at the reason why some documentary standards related to metrological practice (for example [10] and [40]) recommend or even require to take a coverage factor equal to two, irrespective of the application, and without much regard to whether this factor does, in fact, correspond to a coverage probability of 95 %. In many cases the user of the measurement result will need to recover the standard uncertainty from this information in order to apply the LPU.

Chapter 5

The joint treatment of several measurands

The problem addressed in this chapter consists of evaluating the estimates and uncertainties of various output quantities Q_i related to several input quantities Z_j through a number of model relations $\mathcal{M}_k$ that may or may not be explicit. In symbols,

$$\mathcal{M}_1(Q_1, \ldots, Q_{n_Q}, Z_1, \ldots, Z_{n_Z}) = 0$$
$$\vdots \qquad (5.1)$$
$$\mathcal{M}_{n_\mathcal{M}}(Q_1, \ldots, Q_{n_Q}, Z_1, \ldots, Z_{n_Z}) = 0.$$

(Note that not all input and output quantities need to appear in all relations and that the functions $\mathcal{M}_k$ need not be analytic.)

In principle, we could derive the joint pdf $f(q_1, \ldots, q_{n_Q})$ from the joint pdf $f(z_1, \ldots, z_{n_Z})$. The best estimates q_{ei}, the standard uncertainties u_{qi} and the mutual uncertainties $u_{qi,qj}$ would then be obtained by integration. Since in most cases this approach is not feasible in practice, a simplified procedure is needed. If the input best estimates z_{ei} are available, they are inserted in the model relations and we have, for the output best estimates, the system

$$\mathcal{M}_1(q_1, \ldots, q_{n_Q}, z_{e1}, \ldots, z_{en_Z}) = 0$$
$$\vdots \qquad (5.2)$$
$$\mathcal{M}_{n_\mathcal{M}}(q_1, \ldots, q_{n_Q}, z_{e1}, \ldots, z_{en_Z}) = 0$$

which has to be solved by whatever means are possible. Now, this depends on the number of model relations. If $n_\mathcal{M}$ is less than n_Q, the problem is underdetermined: there are simply not enough equations to obtain the output best estimates.

The case $n_\mathcal{M} = n_Q$ will be considered in sections 5.2 and 5.3. It involves solving the system (5.2) and finding the output standard and mutual uncertainties from a generalized form of the LPU in section 3.4. For this, the standard uncertainties u_{zi} and the mutual uncertainties $u_{zi,zj}$ should be available. Examples are presented in section 5.4.

Finally, if $n_M > n_Q$, the problem is overdetermined: there are more equations than unknowns. This means that the values q_{ei} that satisfy the first n_Q model relations will, in general, not satisfy the rest. However, an *adjusted* solution is possible if the estimates of the input quantities are modified in such a way that the resulting values q_{ei}s do satisfy all equations. Since there is no unique way of doing so, an optimality criterion must be used to obtain the 'best' solution. This is provided by the so-called 'least-squares' technique, to which sections 5.5 through 5.8 are devoted.

All equations to be presented in this chapter are best given in compact form by using the matrix formalism, a brief overview of which is provided next.

5.1 Vectors and matrices

A *vector* is just a conventional way of representing with one symbol a collection of related mathematical entities. For example, the set of n values $a_1, \ldots, a_n$ may be written as

$$\mathbf{A} = \begin{pmatrix} a_1 \\ \vdots \\ a_n \end{pmatrix}. \tag{5.3}$$

Now suppose we acquire a first set of n values on a given day, a second set the next day and so forth until the mth day. We might represent all nm values by a vector similar to the one in (5.3), although longer vertically. But if it is convenient or required to keep track of the day on which the values were obtained, it is more appropriate to use a double subscript, and write a_{ij} for the ith value obtained on the jth day. This collection can be represented as

$$\mathbf{A} = \begin{pmatrix} a_{11} & \cdots & a_{1m} \\ \vdots & \ddots & \vdots \\ a_{n1} & \cdots & a_{nm} \end{pmatrix} \tag{5.4}$$

and is referred to as a *matrix* with n rows and m columns, or as an $n \times m$ matrix. We also say that the *dimension* of the matrix is $n \times m$. Note that a vector is equivalent to an $n \times 1$ *column matrix*.

A matrix is *square* if it has the same number of rows and columns. In other words, an $n \times n$ matrix is a square matrix. A square matrix is *symmetric* if its elements are such that $a_{ij} = a_{ji}$. A symmetric matrix is *diagonal* if its elements satisfy the condition $a_{ij} = 0$ for $i \neq j$. The *unit* or *identity* matrix, for which we will reserve the symbol $\mathbf{I}$, is a diagonal matrix whose non-zero elements are all equal to one.

The *transpose* of an $n \times m$ matrix $\mathbf{A}$ is defined as the $m \times n$ matrix $\mathbf{A}^\mathrm{T}$ obtained by interchanging rows and columns. Thus for the matrix in (5.4) its

transpose is
$$\mathbf{A}^T = \begin{pmatrix} a_{11} & \cdots & a_{n1} \\ \vdots & \ddots & \vdots \\ a_{1m} & \cdots & a_{nm} \end{pmatrix}$$
while the transpose of the vector in (5.3) is the *row matrix*
$$\mathbf{A}^T = (a_1 \quad \cdots \quad a_n).$$

Clearly, the transpose of a symmetric matrix is equal to the original matrix. Also, $(\mathbf{A}^T)^T = \mathbf{A}$. Thus, a vector can be written as in (5.3) or, more conveniently within a paragraph, as $\mathbf{A} = (a_1 \quad \cdots \quad a_n)^T$.

An $n \times m$ matrix $\mathbf{A}$ can be *stacked* over an $l \times m$ matrix $\mathbf{B}$ to form an $(n+l) \times m$ matrix $\mathbf{C}$. Similarly, an $n \times m$ matrix $\mathbf{D}$ can be *juxtaposed* to an $n \times l$ matrix $\mathbf{E}$ to form an $n \times (m+l)$ matrix $\mathbf{F}$. Our notation for these matrices will be

$$\mathbf{C} = \begin{pmatrix} \mathbf{A} \\ \mathbf{B} \end{pmatrix} \quad \text{and} \quad \mathbf{F} = (\mathbf{E} \quad \mathbf{D}).$$

These are called *block matrices*. Of course, a block matrix might be formed by stacking and juxtaposing any number of individual matrices, for example

$$\mathbf{E} = \begin{pmatrix} \mathbf{A} & \mathbf{B} \\ \mathbf{C} & \mathbf{D} \end{pmatrix}$$

where the elements in parentheses are matrices of appropriate dimensions.

Two matrices can be added or subtracted only if they are of the same dimension. The rule is

$$\mathbf{A} + \mathbf{B} = \begin{pmatrix} a_{11} + b_{11} & \cdots & a_{n1} + b_{n1} \\ \vdots & \ddots & \vdots \\ a_{1m} + b_{1m} & \cdots & a_{nm} + b_{nm} \end{pmatrix}.$$

Multiplying a matrix (or vector) $\mathbf{A}$ by a constant b results in a new matrix whose elements are those of the original matrix multiplied by the constant:

$$b\mathbf{A} = \begin{pmatrix} ba_{11} & \cdots & ba_{1m} \\ \vdots & \ddots & \vdots \\ ba_{n1} & \cdots & ba_{nm} \end{pmatrix}.$$

An $n \times m$ matrix $\mathbf{A}$ can be multiplied by a matrix $\mathbf{B}$ only if the dimension of the latter is $m \times l$. The result is the $n \times l$ matrix $\mathbf{C} = \mathbf{AB}$ whose elements are

$$c_{ij} = a_{i1}b_{1j} + \cdots + a_{im}b_{mj} \quad i = 1, \ldots, n; \quad j = 1, \ldots, l.$$

In words, the element c_{ij} is obtained as the sum of the products of the elements in row i of $\mathbf{A}$ multiplied by the elements in column j of $\mathbf{B}$. Note that this operation is commutative only if $\mathbf{A}$ and $\mathbf{B}$ are symmetric. However, in general

$$(\mathbf{AB})^T = \mathbf{B}^T \mathbf{A}^T$$

which is left as an exercise for the reader to verify. Also, it is easily proved that the multiplication of an $n \times m$ matrix by an $m \times m$ unit matrix yields the original matrix. Thus, with unit matrices of appropriate dimensions, $\mathbf{IA} = \mathbf{A}$ and $\mathbf{AI} = \mathbf{A}$.

The *inverse* of a square matrix $\mathbf{A}$ is the matrix $\mathbf{A}^{-1}$ such that when multiplied by the original matrix the unit matrix results:

$$\mathbf{AA}^{-1} = \mathbf{A}^{-1}\mathbf{A} = \mathbf{I}.$$

Therefore,
$$\mathbf{AB} = \mathbf{C}$$

is equivalent to
$$\mathbf{A} = \mathbf{CB}^{-1}$$

if $\mathbf{B}$ is square and to
$$\mathbf{B} = \mathbf{A}^{-1}\mathbf{C}$$

if $\mathbf{A}$ is square.

The procedure for obtaining the inverse of a matrix is straightforward, although quite involved if done using a hand calculator for matrices larger than 3×3. The rule will not be given here, since this operation is implemented in most electronic spreadsheets, as well as in more specialized software packages such as Mathcad© and Matlab©. The latter is especially appropriate to program the matrix operations presented in this chapter.

Not all square matrices can be inverted. When this happens it is said that the matrix is *singular*, and means that the rows and columns are not linearly independent. For example, the matrix

$$\begin{pmatrix} 5 & -3 & 1 \\ 8 & 4 & 2 \\ 14 & -4 & 3 \end{pmatrix}$$

is singular, because the third row is obtained by doubling the first row, halving the second row and adding the results.

In the treatment that follows, products of the form

$$\mathbf{C} = \mathbf{ABA}^\mathrm{T} \tag{5.5}$$

will appear often. If $\mathbf{A}$ is an $n \times m$ matrix and $\mathbf{B}$ is an $m \times m$ symmetric matrix, the $n \times n$ matrix $\mathbf{C}$ is symmetric, and is singular if $n > m$.

Occasionally, it will be necessary to add all elements of a matrix. This operation can be represented as

$$\mathbf{1}^\mathrm{T}\mathbf{A}\mathbf{1}$$

where **1** represents a vector with all elements equal to 1. If the dimension of $\mathbf{A}$ is $n \times m$ the **1** on the left is $n \times 1$ and that on the right is $m \times 1$. If $\mathbf{A}$ is a vector, the summation of its elements is

$$\mathbf{1}^\mathrm{T}\mathbf{A}.$$

5.2 The generalized law of propagation of uncertainties (GLPU)

In matrix notation, the set of model relations (5.1) is written compactly as

$$\mathcal{M}(\mathbf{Z}, \mathbf{Q}) = \mathbf{0}$$

where $\mathbf{Z} = (Z_1 \ldots Z_{n_Z})^{\mathrm{T}}$ is the vector of input quantities, $\mathbf{Q} = (Q_1 \ldots Q_{n_Q})^{\mathrm{T}}$ is the vector of output quantities, the symbol $\mathbf{0}$ stands for an $n_{\mathcal{M}} \times 1$ column matrix with all elements equal to zero and $\mathcal{M}(\mathbf{Z}, \mathbf{Q})$ is the column matrix

$$\begin{bmatrix} \mathcal{M}_1(\mathbf{Z}, \mathbf{Q}) \\ \vdots \\ \mathcal{M}_{n_{\mathcal{M}}}(\mathbf{Z}, \mathbf{Q}) \end{bmatrix}.$$

Input information consists of the vector of estimated values $\mathbf{z}_e = (z_{e1} \ldots z_{en_Z})^{\mathrm{T}}$ together with the (symmetric) input *uncertainty* or *covariance* matrix

$$\mathbf{u}_{\mathbf{z}}^2 = \begin{pmatrix} u_{z1}^2 & \cdots & u_{z1, zn_Z} \\ \vdots & \ddots & \vdots \\ u_{z1, zn_Z} & \cdots & u_{zn_Z}^2 \end{pmatrix}.$$

(The off-diagonal elements of this matrix can be expressed alternatively in terms of the input correlation coefficients as defined by equation (3.15).)

In this and the following two sections our attention will be restricted to the case $n_Q = n_{\mathcal{M}}$. The vector of output estimates is then obtained by solving the model relations in terms of the given estimates $\mathbf{z}_e$, i.e. $\mathbf{q}_e$ is the solution of $\mathcal{M}(\mathbf{z}_e, \mathbf{q}) = \mathbf{0}$. Obtaining this solution may become quite involved, however, perhaps requiring some numerical procedure (see section 5.2.1).

To obtain the output uncertainty matrix

$$\mathbf{u}_{\mathbf{q}}^2 = \begin{pmatrix} u_{q1}^2 & \cdots & u_{q1, qn_Q} \\ \vdots & \ddots & \vdots \\ u_{q1, qn_Q} & \cdots & u_{qn_Q}^2 \end{pmatrix}$$

we use a procedure analogous to that in section 3.4. For this we need to define the input and output *sensitivity matrices*

$$\mathbf{S}_{\mathbf{z}} \equiv \begin{pmatrix} \partial \mathcal{M}_1 / \partial Z_1 & \cdots & \partial \mathcal{M}_1 / \partial Z_{n_Z} \\ \vdots & \ddots & \vdots \\ \partial \mathcal{M}_{n_{\mathcal{M}}} / \partial Z_1 & \cdots & \partial \mathcal{M}_{n_{\mathcal{M}}} / \partial Z_{n_Z} \end{pmatrix}$$

and

$$\mathbf{S}_{\mathbf{q}} \equiv \begin{pmatrix} \partial \mathcal{M}_1 / \partial Q_1 & \cdots & \partial \mathcal{M}_1 / \partial Q_{n_Q} \\ \vdots & \ddots & \vdots \\ \partial \mathcal{M}_{n_{\mathcal{M}}} / \partial Q_1 & \cdots & \partial \mathcal{M}_{n_{\mathcal{M}}} / \partial Q_{n_Q} \end{pmatrix}$$

where all derivatives are evaluated at $\mathbf{Z} = \mathbf{z}_e$ and $\mathbf{Q} = \mathbf{q}_e$.

Let us linearize the model relations around the best estimates. This gives

$$\mathcal{M}(\mathbf{Z}, \mathbf{Q}) \approx \mathcal{M}(\mathbf{z}_e, \mathbf{q}_e) + \mathbf{S}_z(\mathbf{Z} - \mathbf{z}_e) + \mathbf{S}_q(\mathbf{Q} - \mathbf{q}_e).$$

But since the best estimates satisfy $\mathcal{M}(\mathbf{z}_e, \mathbf{q}_e) = \mathbf{0}$, then

$$\mathcal{M}(\mathbf{Z}, \mathbf{Q}) \approx \mathbf{S}_z(\mathbf{Z} - \mathbf{z}_e) + \mathbf{S}_q(\mathbf{Q} - \mathbf{q}_e).$$

Therefore, $\mathcal{M}(\mathbf{Z}, \mathbf{Q}) = \mathbf{0}$ is approximately equivalent to

$$\mathbf{S}_q(\mathbf{Q} - \mathbf{q}_e) + \mathbf{S}_z(\mathbf{Z} - \mathbf{z}_e) = \mathbf{0}$$

from which

$$\mathbf{Q} = \mathbf{S}\mathbf{Z} + \mathbf{c}_e \tag{5.6}$$

where

$$\mathbf{c}_e = \mathbf{q}_e - \mathbf{S}\mathbf{z}_e$$

and

$$\mathbf{S} \equiv -(\mathbf{S}_q)^{-1}\mathbf{S}_z.$$

(Note that this is possible only if $n_Q = n_\mathcal{M}$, otherwise $\mathbf{S}_q^{-1}$ is not defined.)

Now recall property (3.6). Application of the multidimensional analogue of that expression to (5.6) yields the 'generalized law of propagation of uncertainties' (GLPU)

$$\mathbf{u}_q^2 = \mathbf{S}\mathbf{u}_z^2\mathbf{S}^T \tag{5.7}$$

valid for linearized model relations that do not deviate much from $\mathcal{M}(\mathbf{Z}, \mathbf{Q})$ for values of $\mathbf{Z}$ and $\mathbf{Q}$ in the neighbourhood of $\mathbf{z}_e$ and $\mathbf{q}_e$, respectively.

Note that, because of the discussion in relation to (5.5), $\mathbf{u}_q^2$ is singular if $n_Z < n_Q$.

5.2.1 The Newton–Raphson method

It was mentioned earlier that obtaining the vector of estimates $\mathbf{q}_e$ from the system of equations $\mathcal{M}(\mathbf{z}_e, \mathbf{q}) = \mathbf{0}$ might require a numerical solution procedure. One such procedure is the *Newton–Raphson* iterative technique. This method is best understood by considering how a root for the one-dimensional function $f(q)$ can be obtained (figure 5.1). Start off with a first guess $q = q_1$, and linearize the function around that point. This gives

$$f(q_1) + f'(q_1)(q - q_1) = 0$$

where

$$f'(q_1) = \left.\frac{df(q)}{dq}\right|_{q=q_1}.$$

The generalized law of propagation of uncertainties (GLPU)

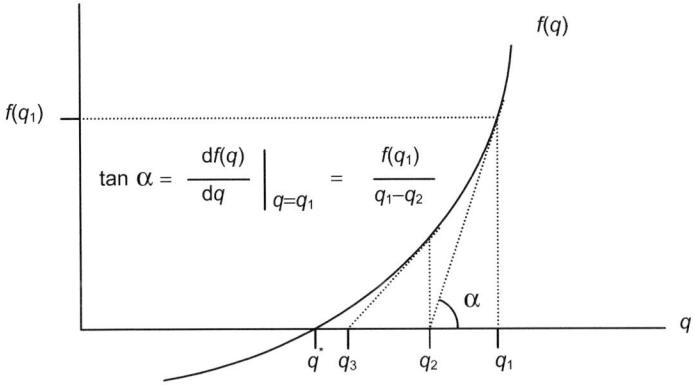

Figure 5.1. The Newton–Raphson iteration in one dimension.

The next guess is then obtained as

$$q_2 = q_1 - \frac{f(q_1)}{f'(q_1)}.$$

If the curve $f(q)$ is 'well behaved' and q_1 is not 'too far' from the root q^*, the iteration should converge quite rapidly.

The multidimensional analogue of this procedure is the following. We write

$$\mathcal{M}(\mathbf{z}_e, \mathbf{q}_1) + \mathbf{S}_{\mathbf{q}_1}(\mathbf{q} - \mathbf{q}_1) = \mathbf{0}$$

from which

$$\mathbf{q}_2 = \mathbf{q}_1 - \left(\mathbf{S}_{\mathbf{q}_1}\right)^{-1} \mathcal{M}(\mathbf{z}_e, \mathbf{q}_1)$$

where $\mathbf{S}_{\mathbf{q}_1}$ means that the partial derivatives in the output sensitivity matrix $\mathbf{S}_{\mathbf{q}}$ are evaluated at $\mathbf{Z} = \mathbf{z}_e$ and $\mathbf{Q} = \mathbf{q}_1$. The convergence values should be close to the required best estimates $\mathbf{q}_e$.

5.2.2 Linear models

Linear models in the output quantities appear frequently. These are of the form

$$\mathcal{M}(\mathbf{Z}) + \mathbf{A}\mathbf{Q} = \mathbf{0}$$

where

$$\mathcal{M}(\mathbf{Z}) \equiv \begin{bmatrix} \mathcal{M}_1(\mathbf{Z}) \\ \vdots \\ \mathcal{M}_{n_\mathcal{M}}(\mathbf{Z}) \end{bmatrix}$$

and $\mathbf{A}$ is a non-singular square constant matrix of dimension $n_\mathcal{M} \times (n_Q = n_\mathcal{M})$. Then $\mathbf{S}_{\mathbf{q}} = \mathbf{A}$ and immediately

$$\mathbf{q}_e = -\mathbf{A}^{-1} \mathcal{M}(\mathbf{z}_e)$$

and
$$\mathbf{u}_q^2 = (\mathbf{A}^{-1}\mathbf{S_z})\mathbf{u}_z^2(\mathbf{A}^{-1}\mathbf{S_z})^\mathrm{T}.$$

The following special cases are of interest:
(i) Models of the form
$$\mathbf{Q} = \mathcal{M}(\mathbf{Z}) \tag{5.8}$$
for which $\mathbf{A} = -\mathbf{I}$. Then
$$\mathbf{q}_e = \mathcal{M}(\mathbf{z}_e)$$
and
$$\mathbf{u}_q^2 = \mathbf{S_z}\mathbf{u}_z^2\mathbf{S_z^T}.$$

When $n_Q = n_\mathcal{M} = 1$ the input sensitivity matrix $\mathbf{S_z}$ turns into a $1 \times n_Z$ row matrix and the expression for u_q^2 reduces to the LPU, equation (3.22).
(ii) Models of the form
$$\mathbf{BZ} + \mathbf{AQ} = 0$$
for which $\mathbf{S_z} = \mathbf{B}$ (an $n_\mathcal{M} \times n_Z$ constant matrix). Then
$$\mathbf{q}_e = -\mathbf{A}^{-1}\mathbf{B}\mathbf{z}_e$$
and
$$\mathbf{u}_q^2 = (\mathbf{A}^{-1}\mathbf{B})\mathbf{u}_z^2(\mathbf{A}^{-1}\mathbf{B})^\mathrm{T}.$$

(iii) Models of the form
$$\mathbf{Q} = \mathbf{BZ}$$
for which $\mathbf{A} = -\mathbf{I}$ and $\mathbf{S_z} = \mathbf{B}$. Then
$$\mathbf{q}_e = \mathbf{B}\mathbf{z}_e$$
and
$$\mathbf{u}_q^2 = \mathbf{B}\mathbf{u}_z^2\mathbf{B}^\mathrm{T}.$$

(iv) Models of the form
$$\mathbf{Z} + \mathbf{AQ} = 0$$
for which $\mathbf{S_z} = \mathbf{I}$ with $n_Z = n_Q = n_\mathcal{M}$. Then
$$\mathbf{q}_e = -\mathbf{A}^{-1}\mathbf{z}_e$$
and
$$\mathbf{u}_q^2 = \mathbf{A}^{-1}\mathbf{u}_z^2(\mathbf{A}^{-1})^\mathrm{T}.$$

5.3 The input uncertainty matrix

To obtain the output uncertainty matrix from the GLPU, equation (5.7), the input uncertainty matrix $\mathbf{u}_z^2$ is needed. If this is not given, it can be obtained from an equation similar to (5.7) if the input quantities $\mathbf{Z}$ are in turn modelled in terms of another set of quantities $\mathbf{Y}$.

By repeated application of this procedure, one will eventually reach a final set of primary quantities whose estimates and standard uncertainties should be obtained as explained in chapter 3. However, in that chapter the problem of determining the mutual uncertainties was not discussed. Of course, these will be zero if it can be assumed that all primary quantities are independent. Often this will indeed be a reasonable assumption. But if a source of correlation between two quantities is suspect, appropriate steps must be taken to evaluate their mutual uncertainty.

One source of correlation, discussed in the *Guide*, is the sequential repeated measurement of two (or more) quantities Z and Y, such that after each value z_i a value y_i follows immediately. It is possible that influence quantities affecting the measurement of both quantities, while varying randomly during the acquisition of all data, will cause increasing or decreasing tendencies on the individual elements of each data pair (z_i, y_i). From an analogy with the TA procedure in section 3.8, in such cases the *Guide* recommends using the formula

$$u_{z,y} = \frac{s_{z,y}}{\sqrt{n}} \qquad (5.9)$$

where

$$s_{z,y}^2 = \frac{1}{n-1} \sum_{k=1}^{n} (z_i - \bar{z})(y_i - \bar{y})$$

and n is the number of measurement pairs.

5.4 Applications of the generalized law of propagation

Three examples on the application of the GLPU are considered in this section. The first is taken from H.2 in the *Guide* and has to do with the simultaneous measurement of two electrical quantities. The second is also taken from the *Guide* (Clause 5.2.2) and involves the calibration of a resistor. Finally, in the third application a method to evaluate comparison measurements is proposed.

Example: measurement of a reactance and a resistance. The resistance X and the reactance Y of a circuit element are to be determined. This is done by measuring the amplitude V of a sinusoidal alternating potential difference across the terminals of the circuit, the amplitude I of the alternating current passing through it, and the phase-shift angle ϕ of the potential difference relative to the current. The

128   The joint treatment of measurands

Table 5.1. Direct measurement data.

Series k	V_k/V	I_k/mA	ϕ_k/rad
1	5.007	19.663	1.0456
2	4.994	19.639	1.0438
3	5.005	19.640	1.0468
4	4.990	19.685	1.0428
5	4.999	19.678	1.0433

model relations are
$$X = Z \cos \phi$$
and
$$Y = Z \sin \phi$$
where
$$Z = \frac{V}{I}.$$

Measurement data are shown in table 5.1. Each series represents simultaneous measurements of the three (primary) input quantities V, I and ϕ.

From a TA analysis of these data (i.e. equations (3.40), (3.41) and (5.9)) we obtain the input estimates, standard uncertainties and correlation coefficients shown in table 5.2. The output estimates are computed directly from the model relations. To obtain the output uncertainty matrix we note that the model relations are of the form (5.8) with input sensitivity matrix

$$\begin{pmatrix} x_e/v_e & -x_e/i_e & -y_e \\ y_e/v_e & -y_e/i_e & x_e \\ z_e/v_e & -z_e/i_e & 0 \end{pmatrix}.$$

Results are shown in table 5.3. Correlation coefficients were calculated through division of the off-diagonal elements of the output uncertainty matrix by the square roots of the corresponding diagonal elements.

Note that (5.9) should be applied only if each series in table 5.1 represents data acquired sequentially. If these data are re-interpreted as representing the five measurements of V, followed by the five measurements of R to end with the five measurements of ϕ, the estimates and standard uncertainties of each of these quantities will be evaluated as before, but all mutual uncertainties should be set to zero. If this is done, the results in table 5.4 are obtained. The best estimates are unchanged, but the output uncertainty matrix is modified drastically.

Table 5.2. Estimates, standard uncertainties and correlation coefficients for the input quantities.

Quantity	Value	Standard uncertainty	Correlation coefficients I	ϕ
V	4.999 00 V	0.003 20 V	-0.355	0.858
I	19.661 00 mA	0.009 47 mA		-0.645
ϕ	1.044 46 rad	0.000 75 rad		

Table 5.3. Estimates, standard uncertainties and correlation coefficients for the output quantities assuming correlated input quantities.

Quantity	Value	Standard uncertainty	Correlation coefficients Y	Z
X	127.73 Ω	0.071 Ω	-0.588	-0.485
Y	219.85 Ω	0.296 Ω		0.993
Z	254.26 Ω	0.236 Ω		

Table 5.4. Estimates, standard uncertainties and correlation coefficients for the output quantities assuming uncorrelated input quantities.

Quantity	Value	Standard uncertainty	Correlation coefficients Y	Z
X	127.73 Ω	0.195 Ω	0.057	0.527
Y	219.85 Ω	0.201 Ω		0.878
Z	254.26 Ω	0.204 Ω		

Example: calibration of an electrical resistor. It is desired to obtain the value and uncertainty of a test resistor T formed by connecting in series n resistors $\mathbf{T} = (T_1 \ \ldots \ T_n)^\mathrm{T}$ of the same nominal value. Assuming all connections have negligible resistance, from the model

$$T = \sum_{i=1}^{n} T_i = \mathbf{1}^\mathrm{T} \mathbf{T}.$$

we get
$$t_e = \sum_{i=1}^{n} t_{ei}$$
and
$$u_t^2 = \mathbf{1}^T \mathbf{u_t^2} \mathbf{1}. \tag{5.10}$$

Data consist of the estimates and associated standard uncertainties of the ratios $\mathbf{R} = (R_1 \ \ldots \ R_n)^T$, where $R_i = T_i/S$ and S is a standard resistor whose best estimate s_e and standard uncertainty u_s are known. The ratios were measured with a comparator bridge such as the one in section 3.11.1.

With the primary quantities written as
$$\mathbf{P} = \begin{pmatrix} S \\ \mathbf{R} \end{pmatrix}$$
the model for the intermediate quantities is
$$\mathbf{T} = \mathcal{M}(\mathbf{P})$$
where
$$\mathcal{M}(\mathbf{P}) = \begin{pmatrix} SR_1 \\ \vdots \\ SR_n \end{pmatrix}.$$
This model is of the form (5.8). Therefore,
$$\mathbf{t_e} = \mathcal{M}(\mathbf{p_e})$$
and
$$\mathbf{u_t^2} = \mathbf{S} \mathbf{u_p^2} \mathbf{S}^T$$
where
$$\mathbf{S} = \begin{pmatrix} r_{e1} & s_e & 0 & \cdots & 0 \\ r_{e2} & 0 & s_e & \cdots & 0 \\ \vdots & \vdots & \vdots & \ddots & \vdots \\ r_{en} & 0 & 0 & \cdots & s_e \end{pmatrix}.$$

Assuming that all estimates of $\mathbf{R}$ are obtained independently, and that their associated standard uncertainties are equal to a common value σ,
$$\mathbf{u_p^2} = \begin{pmatrix} u_s^2 & 0 & \cdots & 0 \\ 0 & \sigma^2 & \cdots & 0 \\ \vdots & \vdots & \ddots & \vdots \\ 0 & 0 & \cdots & \sigma^2 \end{pmatrix}$$

and, therefore,

$$\mathbf{u}_t^2 = \begin{pmatrix} s_e^2\sigma^2 + r_{e1}^2 u_s^2 & r_{e1}r_{e2}u_s^2 & \cdots & r_{e1}r_{en}u_s^2 \\ r_{e1}r_{e2}u_s^2 & s_e^2\sigma^2 + r_{e2}^2 u_s^2 & \cdots & r_{e2}r_{en}u_s^2 \\ \vdots & \vdots & \ddots & \vdots \\ r_{e1}r_{en}u_s^2 & r_{e2}r_{en}u_s^2 & \cdots & s_e^2\sigma^2 + r_{en}^2 u_s^2 \end{pmatrix}$$

or

$$u_{ti}^2 = s_e^2\sigma^2 + r_{ei}^2 u_s^2$$

with correlation coefficients

$$r_{ti,tj} = \frac{r_{ei}r_{ej}u_s^2}{u_{ti}u_{tj}}.$$

Now substitute $\mathbf{u}_t^2$ into (5.10). Since this equation signifies that all elements of the uncertainty matrix $\mathbf{u}_t^2$ are to be added, we get, finally,

$$u_t^2 = ns_e^2\sigma^2 + u_s^2 \sum_{i,j=1}^{n} r_{ei}r_{ej}.$$

If all estimates r_{ei} are equal to one, then

$$t_e = ns_e$$

and

$$u_t^2 = ns_e^2\sigma^2 + n^2 u_s^2$$

with correlation coefficients

$$r_{ti,tj} = \left[1 + \left(\frac{s_e\sigma}{u_s}\right)^2\right]^{-1}.$$

This last expression is given in example 2 of F.1.2.3 of the *Guide*, where numeric values are used to show that $r_{ti,tj} \approx 1$. Thus, even though the comparison ratios are independent, the resistors themselves are nearly perfectly correlated due to their values having been obtained by means of a common measurement standard.

Example: difference comparison measurements. Difference comparison measurements were defined in section 3.11.1 as those by which the value of an attribute T of a test specimen (usually a material measure or a reference material) is established by measuring its difference D with the corresponding attribute S of a reference standard. From the model

$$T = D + S$$

we get
$$t_e = d_e + s_e$$
and
$$u_t^2 = u_d^2 + u_s^2.$$

In order to minimize the influence of random effects, it is advisable to estimate the value of D from redundant comparisons of S and T. Virtually an unlimited number of sequences may be used, for example, $(STS)_n$, $(TST)_n$, $(STTS)_n$, $(TSST)_n$ and so forth, where n is a repeat factor for the cycle in parentheses. In turn, the measured values can be processed following various schemes. For example, two possible schemes for the series of estimates $t_1 s_1 t_2 t_3 s_2 t_4 \ldots$ in the sequence $(TST)_n$ are

$$d_i = \frac{a t_{2i-1} + b t_{2i}}{a + b} - s_i \qquad (5.11)$$

with $i = 1, \ldots, n$ or

$$d_i = \frac{a t_{2i-1} + b t_{2i} + c t_{2i+1} + d t_{2i+2}}{a + b + c + d} - \frac{e s_i + f s_{i+1}}{e + f}$$

with $i = 1, \ldots, n - 1$, where $a, b, \ldots$ are arbitrary constant weight factors. (The subscript e has been suppressed for convenience.)

To evaluate this information, define **P** as the vector of quantities formed by ordering T and S in accordance with the chosen measuring sequence, and **D** as the vector of individual differences estimated by the values d_i. This can be written as

$$\mathbf{D} = \mathbf{A}\mathbf{P}$$

where **A** is a matrix whose form depends on the chosen evaluation scheme.

For example, in the sequence $(TS)_n$ we write

$$\mathbf{P} = (T_1 \quad S_1 \quad T_2 \quad S_2 \quad \ldots)^\mathrm{T}$$

where T_i and S_i denote the ith measurement of T and S, respectively. Then, letting

$$\mathbf{A} = \begin{pmatrix} 0.5 & -1 & 0.5 & 0 & 0 & 0 & \ldots \\ 0 & 0 & 0.5 & -1 & 0.5 & 0 & \ldots \\ 0 & 0 & 0 & 0 & 0.5 & -1 & \ldots \\ \vdots & \vdots & \vdots & \vdots & \vdots & \vdots & \vdots \end{pmatrix} \qquad (5.12)$$

the scheme (5.11) with $a = b = 1$ is obtained:

$$D_i = \frac{T_i + T_{i+1}}{2} - S_i.$$

To evaluate the best estimate for the difference D, one possibility is to calculate the mean of the estimates of the n quantities $\mathbf{D}$. From the model

$$D = \frac{1}{n}\sum_{i=1}^{n} D_i = \mathbf{BP} \qquad (5.13)$$

where

$$\mathbf{B} = \frac{1}{n}\mathbf{1}^{\mathrm{T}}\mathbf{A}$$

we get

$$u_d^2 = \mathbf{B}\mathbf{u}_\mathbf{p}^2\mathbf{B}^{\mathrm{T}}.$$

If it is assumed that the standard uncertainty in the measurement of the input quantities $\mathbf{P}$ is equal to the common value σ, and that these quantities are uncorrelated, then $\mathbf{u}_\mathbf{p}^2 = \sigma^2 \mathbf{I}$ and

$$u_d^2 = \sigma^2 \mathbf{B}\mathbf{B}^{\mathrm{T}}.$$

For example, for $\mathbf{A}$ as in (5.12), it is a simple matter to verify that

$$u_d^2 = \frac{4n-1}{2n^2}\sigma^2.$$

(This equation was derived in [15] using summation notation.)

Note, however, that (5.13) implies giving, in general, different weights to the measured values of S and T. A more rational solution to the problem of evaluating comparison measurements is given in section 5.8.3.

5.5 Least-squares adjustment

In this section we address the case of $n_\mathcal{M}$ independent model relations $\mathcal{M}(\mathbf{Z}, \mathbf{Q}) = \mathbf{0}$ for n_Z input quantities $\mathbf{Z}$ and n_Q output quantities $\mathbf{Q}$, where $n_Q < n_\mathcal{M}$. We are given the vector of estimates $\mathbf{z}_e$ and the uncertainty matrix $\mathbf{u}_\mathbf{z}^2$, and we wish to obtain the vector $\mathbf{q}_e$ and the uncertainty matrix $\mathbf{u}_\mathbf{q}^2$.

Example. Let

$$Q = 5Z_1 - 3Z_2$$

and

$$Q = 4Z_1 + Z_2.$$

Thus, $n_Q = 1 < n_\mathcal{M} = n_Z = 2$. Assume $z_{e1} = 5$ and $z_{e2} = 1$. Then $q_e = 22$ results from the first relation and $q_e = 21$ results from the second one. A common value for q_e would result only by forcing $5Z_1 - 3Z_2$ to be equal to $4Z_1 + Z_2$, i.e. by having input estimates satisfying $z_{e1} = 4z_{e2}$. For example $q_e = 17$ obtains if $z_{e1} = 4$ and $z_{e2} = 1$.

The solution to this problem is approached by seeking *adjusted* estimates of the input quantities, call them $\mathbf{z}_a$, such that when introduced into the model relations only one vector $\mathbf{q}_e$ is obtained for the output quantities. In other words, the vectors $\mathbf{z}_a$ and $\mathbf{q}_e$ comply with $\mathcal{M}(\mathbf{z}_a, \mathbf{q}_e) = \mathbf{0}$. In general, however, $\mathbf{z}_a$ is not unique. In the previous example, any value v could be chosen such that $z_{a1} = 4v$ and $z_{a2} = v$, from which $q_e = 17v$ would result.

Before proceeding to discuss the solution to the adjustment problem, it should be noted that, besides having $n_Q < n_\mathcal{M}$, the inequality $n_\mathcal{M} \leq n_Z$ must also hold. Otherwise it would, in general, be impossible to find adjusted values for $\mathbf{Z}$ yielding a unique vector of estimates $\mathbf{q}_e$.

Example. Suppose that a third linearly independent relation

$$Q = 3Z_1 + 5Z_2$$

is added to those in the previous example. Clearly, no value q_e will now satisfy all three relations, whatever the values of the input quantities.

5.5.1 Classical least-squares

Intuitively, the 'best' adjusted values $\mathbf{z}_a$ should be those that deviate the least from the data $\mathbf{z}_e$. In the classical least-squares procedure, devised by Gauss, the values $\mathbf{z}_a$ are obtained by minimizing the sum

$$\chi_c^2 = \sum_{i=1}^{n_Z}(z_i - z_{ei})^2 = (\mathbf{z} - \mathbf{z}_e)^T(\mathbf{z} - \mathbf{z}_e) \tag{5.14}$$

with respect to $\mathbf{z}$, subject to $\mathcal{M}(\mathbf{z}, \mathbf{q}) = \mathbf{0}$. Note that this criterion is reasonable, but arbitrary. For example, one might as well seek a minimum for the sum of the absolute deviations $|z_i - z_{ei}|$.

When applied to measurement data, minimizing χ_c^2 presents at least three inconveniences. First, if the dimensions (or units) of the summands in (5.14) are different, χ_c^2 is devoid of physical meaning. Second, the classical procedure does not take into account the uncertainty associated with the estimates of the differences whose squared sum is being minimized. (This fact would make it questionable to compute *formally* the uncertainty matrix $\mathbf{u}_\mathbf{q}^2$ using (5.7) with $\mathbf{S}$ evaluated at $\mathbf{z}_a$ and $\mathbf{q}_e$.) Finally, the classical procedure does not allow one to verify whether the solution is *consistent* with the data or the model (we return to this point in section 5.7).

5.5.2 Least-squares under uncertainty

As presented in [9,51,53], the least-squares adjustment method under uncertainty proceeds by minimizing

$$\chi^2 = (\mathbf{z} - \mathbf{z}_e)^T (\mathbf{u}_\mathbf{z}^2)^{-1} (\mathbf{z} - \mathbf{z}_e) \tag{5.15}$$

with respect to $\mathbf{z}$, subject to $\mathcal{M}(\mathbf{z}, \mathbf{q}) = \mathbf{0}$. This expression is adimensional; it reduces to

$$\chi^2 = \sum_{i=1}^{n_Z} \left(\frac{z_i - z_{ei}}{u_{zi}} \right)^2$$

if the input uncertainty matrix $\mathbf{u}_\mathbf{z}^2$ is diagonal.

Note that minimizing χ_c^2 is equivalent to minimizing χ^2 if the input uncertainty matrix is equal to the identity matrix. Hence, the classical solution follows from the following equations by setting formally $\mathbf{u}_\mathbf{z}^2 = \mathbf{I}$.

The minimization of (5.15) is done by introducing a vector $\mathbf{L}$ of $n_\mathcal{M}$ unknown *Lagrange multipliers* and minimizing instead the modified sum

$$\chi_m^2 = (\mathbf{z} - \mathbf{z}_e)^T (\mathbf{u}_\mathbf{z}^2)^{-1} (\mathbf{z} - \mathbf{z}_e) + 2\mathcal{M}(\mathbf{z}, \mathbf{q})^T \mathbf{l}$$

with respect to both $\mathbf{z}$ and $\mathbf{q}$, where $\mathbf{l}$ represents the values of $\mathbf{L}$ (not to be confused with the 'one' vector $\mathbf{1}$). The solution yields the values $\mathbf{z}_a$ and $\mathbf{q}_e$ satisfying $\mathcal{M}(\mathbf{z}_a, \mathbf{q}_e) = \mathbf{0}$. Therefore, the minimum values of χ^2 and χ_m^2 will coincide.

We now differentiate $\chi_m^2/2$ with respect to $\mathbf{z}$ and $\mathbf{q}$, setting the results equal to zero. This gives

$$(\mathbf{u}_\mathbf{z}^2)^{-1} (\mathbf{z} - \mathbf{z}_e) + \mathbf{S}_\mathbf{z}(\mathbf{z}, \mathbf{q})^T \mathbf{l} = \mathbf{0} \tag{5.16}$$

and

$$\mathbf{S}_\mathbf{q}(\mathbf{z}, \mathbf{q})^T \mathbf{l} = \mathbf{0} \tag{5.17}$$

where the somewhat cumbersome notation $\mathbf{S}_\mathbf{z}(\mathbf{z}, \mathbf{q})$ and $\mathbf{S}_\mathbf{q}(\mathbf{z}, \mathbf{q})$ is needed to specify that the derivatives in the input and output sensitivity matrices $\mathbf{S}_\mathbf{z}$ and $\mathbf{S}_\mathbf{q}$ are to be left in terms of $\mathbf{z}$ and $\mathbf{q}$. Considering also the model relations

$$\mathcal{M}(\mathbf{z}, \mathbf{q}) = \mathbf{0}$$

we end up with a system of $n_Z + n_Q + n_\mathcal{M}$ equations for the $n_Z + n_Q + n_\mathcal{M}$ unknowns $\mathbf{z}$, $\mathbf{q}$ and $\mathbf{l}$.

Since this system is in general nonlinear, an approximate solution is achieved by linearizing the model relations around the given estimates $\mathbf{z}_e$. This gives

$$\mathcal{M}(\mathbf{z}_e, \mathbf{q}) + \mathbf{S}_\mathbf{z}(\mathbf{z}_e, \mathbf{q})(\mathbf{z} - \mathbf{z}_e) = \mathbf{0} \tag{5.18}$$

where $\mathbf{S}_\mathbf{z}(\mathbf{z}_e, \mathbf{q})$ indicates that now the input sensitivity matrix $\mathbf{S}_\mathbf{z}$ is to be evaluated at $\mathbf{Z} = \mathbf{z}_e$ while the values $\mathbf{q}$ are left as variables. To be consistent, the sensitivity matrices in (5.16) and (5.17) should also be evaluated likewise. This gives

$$(\mathbf{u}_\mathbf{z}^2)^{-1} (\mathbf{z} - \mathbf{z}_e) + \mathbf{S}_\mathbf{z}(\mathbf{z}_e, \mathbf{q})^T \mathbf{l} = \mathbf{0} \tag{5.19}$$

and
$$\mathbf{S_q}(\mathbf{z_e}, \mathbf{q})^T \mathbf{l} = \mathbf{0}. \tag{5.20}$$

The system (5.18) + (5.19) + (5.20) is now linear in $\mathbf{l}$ and $\mathbf{z}$, therefore these two vectors can be eliminated. Thus from (5.19),

$$\mathbf{z} - \mathbf{z_e} = -\mathbf{u}_z^2 \mathbf{S_z}(\mathbf{z_e}, \mathbf{q})^T \mathbf{l}. \tag{5.21}$$

Now substitute this expression into (5.18). This gives

$$\mathcal{M}(\mathbf{z_e}, \mathbf{q}) - \mathbf{S_z}(\mathbf{z_e}, \mathbf{q})\mathbf{u}_z^2 \mathbf{S_z}(\mathbf{z_e}, \mathbf{q})^T \mathbf{l} = \mathbf{0}$$

from which

$$\mathbf{l} = \mathbf{K}(\mathbf{q})\mathcal{M}(\mathbf{z_e}, \mathbf{q}) \tag{5.22}$$

where

$$\mathbf{K}(\mathbf{q}) \equiv [\mathbf{S_z}(\mathbf{z_e}, \mathbf{q})\mathbf{u}_z^2 \mathbf{S_z}(\mathbf{z_e}, \mathbf{q})^T]^{-1}.$$

Finally, substitution of (5.22) into (5.20) yields a system of n_Q equations for the variables $\mathbf{q}$:

$$\mathbf{S_q}(\mathbf{z_e}, \mathbf{q})^T \mathbf{K}(\mathbf{q}) \mathcal{M}(\mathbf{z_e}, \mathbf{q}) = \mathbf{0} \tag{5.23}$$

whose solution is the vector of estimates of the output quantities, $\mathbf{q_e}$.

If the system (5.23) is nonlinear, the Newton–Raphson algorithm described in section 5.2.1 may be used to solve it. With a suitable starting point $\mathbf{q}_1$, this algorithm is

$$\mathbf{q}_{n+1} = \mathbf{q}_n - [\mathbf{R_q}(\mathbf{q}_n)]^{-1} \mathbf{R}(\mathbf{q}_n) \tag{5.24}$$

where n denotes the iteration step,

$$\mathbf{R}(\mathbf{q}) \equiv \mathbf{S_q}(\mathbf{z_e}, \mathbf{q})^T \mathbf{K}(\mathbf{q}) \mathcal{M}(\mathbf{z_e}, \mathbf{q})$$

and

$$\mathbf{R_q}(\mathbf{q}) \equiv \mathbf{S_q}(\mathbf{z_e}, \mathbf{q})^T \mathbf{K}(\mathbf{q}) \mathbf{S_q}(\mathbf{z_e}, \mathbf{q})$$

(the derivatives of $\mathbf{S_q}(\mathbf{z_e}, \mathbf{q})$ and of $\mathbf{K}(\mathbf{q})$ with respect to $\mathbf{q}$ have been neglected).

Once the vector $\mathbf{q_e}$ is available, one computes the output uncertainty matrix as

$$\mathbf{u}_q^2 = [\mathbf{S_q}(\mathbf{z_e}, \mathbf{q_e})^T \mathbf{K}(\mathbf{q_e}) \mathbf{S_q}(\mathbf{z_e}, \mathbf{q_e})]^{-1} \tag{5.25}$$

and finds the adjusted values by combining (5.21) and (5.22). This yields

$$\mathbf{z_a} = \mathbf{z_e} - \mathbf{u}_z^2 \mathbf{S_z}(\mathbf{z_e}, \mathbf{q_e})^T \mathbf{K}(\mathbf{q_e}) \mathcal{M}(\mathbf{z_e}, \mathbf{q_e}). \tag{5.26}$$

Reaffirming what was mentioned in the introduction to this section, we note that least-squares adjusted values can be found only if

$$n_Q < n_M \leq n_Z. \tag{5.27}$$

Otherwise the matrix

$$\mathbf{S_z}(\mathbf{z_e}, \mathbf{q}) \mathbf{u}_z^2 \mathbf{S_z}(\mathbf{z_e}, \mathbf{q})^T$$

would be singular and it would not be possible to construct the matrix $\mathbf{K}(\mathbf{q})$. (Recall the discussion in relation to (5.5).)

5.5.3 Linear models

As discussed in section 5.2.2, linear models in the output quantities are of the form
$$\mathcal{M}(\mathbf{Z}) + \mathbf{AQ} = \mathbf{0} \tag{5.28}$$
where $\mathcal{M}(\mathbf{Z})$ is the column matrix
$$\begin{bmatrix} \mathcal{M}_1(\mathbf{Z}) \\ \vdots \\ \mathcal{M}_{n_\mathcal{M}}(\mathbf{Z}) \end{bmatrix}$$
and $\mathbf{A}$ is an $n_\mathcal{M} \times n_Q$ constant matrix. In this case $\mathbf{S_q} = \mathbf{A}$ and $\mathbf{S_z}$ does not depend on $\mathbf{q}$. Then, with
$$\mathbf{K} = [\mathbf{S_z}(\mathbf{z_e})\mathbf{u}_z^2 \mathbf{S_z}(\mathbf{z_e})^T]^{-1}$$
the system (5.23) reduces to
$$\mathbf{A}^T \mathbf{K}[\mathcal{M}(\mathbf{z_e}) + \mathbf{Aq}] = \mathbf{0}$$
whose solution is
$$\mathbf{q_e} = -(\mathbf{A}^T \mathbf{KA})^{-1} \mathbf{A}^T \mathbf{K} \mathcal{M}(\mathbf{z_e})$$
$$\mathbf{z_a} = \mathbf{z_e} - \mathbf{u}_z^2 \mathbf{S_z}(\mathbf{z_e})^T \mathbf{K}[\mathcal{M}(\mathbf{z_e}) + \mathbf{Aq_e}]$$
and
$$\mathbf{u_q^2} = (\mathbf{A}^T \mathbf{KA})^{-1}.$$

The following special cases are of interest:

(i) Models of the form
$$\mathbf{Q} = \mathcal{M}(\mathbf{Z})$$
for which $\mathbf{A} = -\mathbf{I}$. Then
$$\mathbf{q_e} = \mathcal{M}(\mathbf{z_e})$$
$$\mathbf{z_a} = \mathbf{z_e} - \mathbf{u}_z^2 \mathbf{S_z}(\mathbf{z_e})^T \mathbf{K}[\mathcal{M}(\mathbf{z_e}) - \mathbf{q_e}]$$
and
$$\mathbf{u_q^2} = \mathbf{S_z}(\mathbf{z_e})\mathbf{u}_z^2 \mathbf{S_z}(\mathbf{z_e})^T.$$

(ii) Models of the form
$$\mathbf{BZ} + \mathbf{AQ} = \mathbf{0} \tag{5.29}$$
for which $\mathbf{S_z} = \mathbf{B}$ is an $n_\mathcal{M} \times n_Z$ constant matrix. Then, with $\mathbf{K} = (\mathbf{Bu}_z^2\mathbf{B}^T)^{-1}$,
$$\mathbf{q_e} = -(\mathbf{A}^T \mathbf{KA})^{-1} \mathbf{A}^T \mathbf{KBz_e}$$
$$\mathbf{z_a} = \mathbf{z_e} - \mathbf{u}_z^2 \mathbf{B}^T \mathbf{K}(\mathbf{Bz_e} + \mathbf{Aq_e})$$

and
$$u_q^2 = (A^T K A)^{-1}.$$

(iii) Models of the form
$$Q = BZ \tag{5.30}$$
for which $A = -I$, $S_z = B$ and $K = (Bu_z^2 B^T)^{-1}$. Then
$$q_e = B z_e$$
$$z_a = z_e$$
and
$$u_q^2 = B u_z^2 B^T.$$

(iv) Models of the form
$$Z + AQ = 0 \tag{5.31}$$
for which $n_\mathcal{M} = n_Z$, $S_z = I$ and $K = (u_z^2)^{-1}$. Then
$$q_e = -[A^T (u_z^2)^{-1} A]^{-1} A^T (u_z^2)^{-1} z_e$$
$$z_a = -A q_e$$
and
$$u_q^2 = [A^T (u_z^2)^{-1} A]^{-1}.$$

5.6 Curve-fitting

Curve-fitting is a special case of least-squares adjustment. It concerns passing a curve through a two-dimensional set of points on a coordinate plane. The form of the curve is fixed and its parameters are to be determined. For example, if the points are deemed to fall along a straight line, the parameters are its slope and intercept. The following discussion is taken from [34].

Consider two quantities X and Y that vary according to the measurement conditions, represented by τ. Let these conditions be varied discretely and define the vectors $\mathbf{X}$ and $\mathbf{Y}$ formed by the quantities $X_i = X(\tau_i)$ and $Y_i = Y(\tau_i)$, with $i = 1, \ldots, I$. These quantities may be thought of as corresponding to the horizontal and vertical coordinates, respectively, of a set of points in the Cartesian plane (figure 5.2). The estimated values $\mathbf{x}_e$ and $\mathbf{y}_e$ and the uncertainty matrices $\mathbf{u}_x^2$ and $\mathbf{u}_y^2$ are given. We assume that $\mathbf{X}$ and $\mathbf{Y}$ are uncorrelated, and that the values $\mathbf{x}_e$ and $\mathbf{y}_e$ have been corrected for systematic effects.

Now let $\mathbf{f}(X)$ be a vector of J independent functions of X, with $J \leq I$. These functions are all given. It is desired to infer the values of the J coefficients $\mathbf{Q}$ in the adjustment curve
$$Y = \mathbf{f}(X)^T \mathbf{Q} = f_1(X) Q_1 + \cdots + f_J(X) Q_J.$$

Curve-fitting

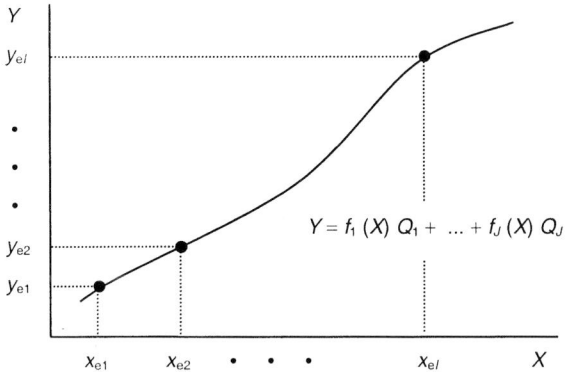

Figure 5.2. A given set of points through which passes a curve.

Note that the curve is linear in the coefficients **Q** but is, in general, nonlinear in X. For example, if the curve is a polynomial

$$Y = Q_1 + Q_2 X + \cdots + Q_J X^{J-1}$$

then

$$\mathbf{f}(X) = (1 \quad X \quad \ldots \quad X^{J-1})^{\mathrm{T}}.$$

5.6.1 Solution using the generalized formalism

To use the generalized formalism in section 5.5.2, let us write **Z** as a vector formed by stacking **X** over **Y**, i.e.

$$\mathbf{Z} = \begin{pmatrix} \mathbf{X} \\ \mathbf{Y} \end{pmatrix}. \tag{5.32}$$

We then have $n_Z = 2I$ input quantities and $n_Q = J < n_Z$ output quantities. The $n_\mathcal{M} = I$ model relations are

$$\mathbf{F}(\mathbf{X})\mathbf{Q} - \mathbf{Y} = \mathbf{0}$$

where

$$\mathbf{F}(\mathbf{X}) = \begin{bmatrix} f_1(X_1) & \cdots & f_J(X_1) \\ \vdots & \ddots & \vdots \\ f_1(X_I) & \cdots & f_J(X_I) \end{bmatrix}.$$

Further, since we have assumed that **X** and **Y** are uncorrelated,

$$\mathbf{u}_\mathbf{z}^2 = \begin{pmatrix} \mathbf{u}_\mathbf{x}^2 & 0 \\ 0 & \mathbf{u}_\mathbf{y}^2 \end{pmatrix} \tag{5.33}$$

where the zeros stand for $I \times I$ matrices with all its elements equal to zero.

The sensitivity matrices are

$$\mathbf{S_q(x) = F(x)}$$

and

$$\mathbf{S_z(x, q) = [D(x, q) \quad -I]}$$

where $\mathbf{D(x, q)}$ is the $I \times I$ diagonal matrix with elements

$$D_{ii} = f'_1(x_i)q_1 + \cdots + f'_J(x_i)q_J$$

where

$$f'_j(x_i) \equiv \left.\frac{\mathrm{d}f_j(X)}{\mathrm{d}X}\right|_{X=x_i}.$$

The estimates $\mathbf{q_e}$ are found by solving

$$\mathbf{F(x_e)^T K(q)[F(x_e)q - y_e] = 0} \tag{5.34}$$

which is the equivalent of (5.23). In this case

$$\mathbf{K(q) = [D(x_e, q)u_x^2 D(x_e, q) + u_y^2]^{-1}}.$$

The solution to (5.34) may be found using the Newton–Raphson algorithm (5.24) with (5.36) as the starting point, where in this case

$$\mathbf{R(q) \equiv F(x_e)^T K(q)[F(x_e)q - y_e]}$$

and

$$\mathbf{R_q(q) \equiv F(x_e)^T K(q) F(x_e)}.$$

After finding the vector $\mathbf{q_e}$, (5.25) gives the output uncertainty matrix

$$\mathbf{u_q^2 = [F(x_e)^T K(q_e) F(x_e)]} \tag{5.35}$$

while the adjusted values are obtained from (5.26):

$$\mathbf{x_a = x_e - u_x^2 D(x_e, q_e) K(q_e)[F(x_e)q_e - y_e]}$$
$$\mathbf{y_a = y_e + u_y^2 K(q_e)[F(x_e)q_e - y_e]}.$$

Note that the problem is very simple when there is no uncertainty associated with the estimates of $\mathbf{X}$, because then these quantities are treated as perfectly known and the model reduces to

$$\mathbf{F(x_e)Q - Y = 0}$$

where the only input quantities are $\mathbf{Y}$. This model is of the form (5.31) with $\mathbf{A = -F(x_e)}$. Accordingly,

$$\mathbf{q_e = [F(x_e)^T (u_y^2)^{-1} F(x_e)]^{-1} F(x_e)^T (u_y^2)^{-1} y_e} \tag{5.36}$$
$$\mathbf{y_a = F(x_e) q_e}$$

and
$$u_q^2 = [F(x_e)^T(u_y^2)^{-1}F(x_e)]^{-1}.$$

Finally, note that if no uncertainties are associated with either coordinate axis, we can formally take $u_y^2 = I$ in (5.36). This yields

$$q_e = [F(x_e)^T F(x_e)]^{-1} F(x_e)^T y_e$$

and
$$y_a = F(x_e)q_e$$

with no uncertainty associated with q_e. This is the classical solution to the curve adjustment problem.

Example. The problem of straight-line fitting

$$Y = Q_1 + Q_2 X$$

arises frequently. In this case we have

$$f(X) = (1 \quad X)^T$$

$$F(X) = \begin{pmatrix} 1 & X_1 \\ \vdots & \vdots \\ 1 & X_I \end{pmatrix}$$

and
$$D(x, q) = q_2 I$$

where I is the identity matrix.

However, system (5.34) is, in this case, nonlinear and has to be solved numerically. An analytic solution is possible if the input uncertainty matrices are of the form $u_x^2 = u_x^2 I$ and $u_y^2 = u_y^2 I$, in which case we have

$$q_e = \frac{1}{D}\begin{pmatrix} S_{xx}S_y - S_x S_{xy} \\ I S_{xy} - S_x S_y \end{pmatrix}$$

and
$$u_q^2 = \frac{q_{e2}^2 u_x^2 + u_y^2}{D}\begin{pmatrix} S_{xx} & -S_x \\ -S_x & I \end{pmatrix}$$

where

$$S_x = \sum_{i=1}^{I} x_{ei}$$

$$S_y = \sum_{i=1}^{I} y_{ei}$$

Table 5.5. Data for the calibration of a thermometer by means of an SPRT.

Point	$x_{ei}/°C$	u_{xi}/mK	$y_{ei}/°C$	u_{yi}/mK
1	0.0010	17.5	0.1335	10.6
2	10.0867	18.6	9.9303	10.7
3	20.2531	18.3	19.8774	10.8
4	30.3477	18.6	29.8626	11.0
5	40.6713	18.8	39.8619	11.3
6	50.8418	19.1	49.8241	11.5
7	60.9435	20.2	59.7264	9.8
8	71.1146	20.4	69.6918	10.1
9	81.3378	21.0	79.6740	10.4
10	91.5948	21.7	89.7215	10.7

$$S_{xx} = \sum_{i=1}^{I} x_{ei}^2$$

$$S_{xy} = \sum_{i=1}^{I} x_{ei} y_{ei}$$

and

$$D = I S_{xx} - s_x^2.$$

As an illustration, consider the following application, taken from [37]. It deals with the calibration of a digital thermometer in the range 0–90 °C by means of a standard platinum resistance thermometer (SPRT) connected to a digital precision multimeter. Calibration data, shown in table 5.5, lead to $q_{e1} = 0.0991$ °C, $q_{e2} = 0.978\,39$, $u_{q1} = 0.0123$ °C, $u_{q2} = 0.000\,24$ and $u_{q1,q2} = -2.41 \times 10^{-6}$ °C. If the input uncertainties are modified to $u_x = 20$ mK and $u_y = 10$ mK for all calibrated points, the new results are $q_{e1} = 0.0979$ °C, $q_{e2} = 0.978\,42$, $u_{q1} = 0.0129$ °C, $u_{q2} = 0.000\,24$ and $u_{q1,q2} = -2.58 \times 10^{-6}$ °C.

5.6.2 An alternative solution

An alternative procedure for minimizing χ^2 is presented next. We start by noting that, with (5.32) and (5.33), equation (5.15) becomes

$$\chi^2 = (\mathbf{x} - \mathbf{x}_e)^T (\mathbf{u}_\mathbf{x}^2)^{-1} (\mathbf{x} - \mathbf{x}_e) + (\mathbf{y} - \mathbf{y}_e)^T (\mathbf{u}_\mathbf{y}^2)^{-1} (\mathbf{y} - \mathbf{y}_e). \qquad (5.37)$$

Curve-fitting

This expression can be minimized directly, subject to

$$\mathbf{y} = \mathbf{F(x)q} \tag{5.38}$$

without recourse to Lagrange multipliers. For this, we substitute (5.38) into (5.37). This gives an unconstrained expression for χ^2 as a function of $\mathbf{x}$ and $\mathbf{q}$ only, which can be minimized by use of the Levenberg–Marquardt algorithms (see e.g. [14]). Here, we proceed to differentiate this expression with respect to both $\mathbf{x}$ and $\mathbf{q}$ and set the results equal to zero. This gives

$$(\mathbf{u}_\mathbf{x}^2)^{-1}(\mathbf{x} - \mathbf{x}_e) + \mathbf{D(x, q)}(\mathbf{u}_\mathbf{y}^2)^{-1}[\mathbf{F(x)q} - \mathbf{y}_e] = 0 \tag{5.39}$$

and

$$\mathbf{F(x)}^T (\mathbf{u}_\mathbf{y}^2)^{-1}[\mathbf{F(x)q} - \mathbf{y}_e] = 0. \tag{5.40}$$

Solving for $\mathbf{q}$ in (5.40) gives

$$\mathbf{q} = [\mathbf{F(x)}^T (\mathbf{u}_\mathbf{y}^2)^{-1} \mathbf{F(x)}]^{-1} \mathbf{F(x)}^T (\mathbf{u}_\mathbf{y}^2)^{-1} \mathbf{y}_e.$$

Substitution of this expression into (5.39) produces a system of equations in $\mathbf{x}$ only, whose solution yields the adjusted values $\mathbf{x}_a$. For this, we again use the Newton–Raphson algorithm, giving

$$\mathbf{x}_{n+1} = \mathbf{x}_n - [\mathbf{R}_\mathbf{x}(\mathbf{x}_n)]^{-1} \mathbf{R}(\mathbf{x}_n)$$

where

$$\mathbf{R}(\mathbf{x}_n) = (\mathbf{u}_\mathbf{x}^2)^{-1}(\mathbf{x}_n - \mathbf{x}_e) + \mathbf{D}(\mathbf{x}_n, \mathbf{q}_n)(\mathbf{u}_\mathbf{y}^2)^{-1}[\mathbf{F}(\mathbf{x}_n)\mathbf{q}_n - \mathbf{y}_e]$$
$$\mathbf{R}_\mathbf{x}(\mathbf{x}_n) = (\mathbf{u}_\mathbf{x}^2)^{-1} + \mathbf{D}(\mathbf{x}_n, \mathbf{q}_n)(\mathbf{u}_\mathbf{y}^2)^{-1} \mathbf{D}(\mathbf{x}_n, \mathbf{q}_n)$$

and

$$\mathbf{q}_n = [\mathbf{F}(\mathbf{x}_n)^T (\mathbf{u}_\mathbf{y}^2)^{-1} \mathbf{F}(\mathbf{x}_n)]^{-1} \mathbf{F}(\mathbf{x}_n)^T (\mathbf{u}_\mathbf{y}^2)^{-1} \mathbf{y}_e.$$

The algorithm starts with $\mathbf{x}_1 = \mathbf{x}_e$, while no starting values for $\mathbf{q}$ are needed. This advantage is achieved at the expense of a very slow convergence. The convergence values $\mathbf{x}_n$ constitute the adjusted values $\mathbf{x}_a$, while the convergence values $\mathbf{q}_n$ give the estimates $\mathbf{q}_e$. Then

$$\mathbf{y}_a = \mathbf{F}(\mathbf{x}_a)\mathbf{q}_e.$$

The uncertainty matrix of the coefficients is again given by (5.35).

(It is interesting to mention that the method in [46] is equivalent to the Newton–Raphson procedure in section 5.6.1. However, the algorithm in [46] was derived starting from equations (5.39) and (5.40).)

5.7 Consistency analysis

After having done a least-squares adjustment, it is important to verify whether the adjusted values conform with the given data. By this, we mean the following. Consider the sum

$$\chi^2 = (\mathbf{Z} - \mathbf{z}_e)^T (\mathbf{u}_Z^2)^{-1} (\mathbf{Z} - \mathbf{z}_e).$$

From probability theory, we know that if $\mathbf{Z}$ follows a multivariate normal pdf with expectation $\mathbf{z}_e$ and variance matrix $\mathbf{u}_Z^2$, then χ^2 follows a chi-squared pdf with $\nu = n_Z$ degrees of freedom.

In our case, the minimum value of χ^2, which we will denote as $\chi^2_{\min}$, serves to determine $n_Z + n_Q$ values from n_Z input estimates subject to $n_\mathcal{M}$ model relations, or *constraints*. Therefore, χ^2 will approximately follow a chi-squared pdf with $\nu = n_Z + n_\mathcal{M} - (n_Z + n_Q) = n_\mathcal{M} - n_Q$ degrees of freedom. Since the expectation of a chi-squared pdf is equal to ν, we expect $\chi^2_{\min}$ to be close to $n_\mathcal{M} - n_Q$ or, equivalently, the so-called *Birge ratio*†

$$\text{Bi} = \left(\frac{\chi^2_{\min}}{n_\mathcal{M} - n_Q} \right)^{1/2}$$

to be approximately equal to one.

Of course, if this happens it cannot be taken as proof that the model is correct. However, a Birge ratio differing significantly from unity is an indication that something is amiss.

The case $\text{Bi} < 1$ suggests that χ^2 is 'too small', meaning that the uncertainty matrix $\mathbf{u}_Z^2$ is somewhat larger than reasonable. In this case it might prove convenient to try to gather more data, if at all possible, or else to investigate whether some of the input uncertainties can be reduced.

The contrary case is a Birge ratio larger than one. Its significance can be investigated by recalling that the variance of a chi-squared pdf with ν degrees of freedom is equal to 2ν. The normalized error

$$e_n = \frac{\chi^2_{\min} - \nu}{(2\nu)^{1/2}}$$

can then be computed with $\nu = n_\mathcal{M} - n_Q$ and compared with a value in the range from one to three [54]. If e_n is larger, it implies that the adjusted values $\mathbf{z}_a$ are 'too far' from the data $\mathbf{z}_e$. This may occur because the model is not correct. For example, a previously unsuspected systematic effect may have been overlooked or the degree of the adjustment polynomial in curve-fitting may be too low. Measurement outliers are also a cause of an unreasonably large value of $\chi^2_{\min}$.

† R T Birge was in charge of coordinating the first evaluation of the best values of the fundamental physical constants in the late 1920s.

However, if it is not possible to revise the model (e.g. because it is fixed by physical considerations), and if no mistakes in the data are apparent, one might think of *enlarging* the input uncertainties so as to force consistency. This can be done by multiplying the matrix $\mathbf{u}_z^2$ by a factor larger than one, in which case the values q_e are not modified but the uncertainty $\mathbf{u}_q^2$ is increased. If this factor is chosen to be the square of the Birge ratio, the new Birge ratio becomes equal to unity.

Alternatively, if $\mathbf{u}_z^2$ is diagonal, a 'cut-off' value can be defined, such that all elements below the cut-off are reset to that value. The value of the cut-off can be decided, for example, on the basis of what is thought to be the state of the art for the particular measurands or such that the new Birge ratio equals one. Similarly, a fixed 'add-on' uncertainty value can be established. Neither of these methods, however, can be justified theoretically. Recently, a method to remove data or model non-conformity based on the Bayesian theory has been proposed [54].

Example. Suppose we are to determine an estimate of Q from the relations $Q = Z_1$ and $Q = Z_2$. This is a model of the form (5.30) with $\mathbf{B} = (1 \quad 1)^T$. Let $z_{e1} = 10$ and $z_{e2} = 14$ with $u_{z1} = 1$ and $u_{z2} = 1.2$ (arbitrary units); the values $q_e = 11.64$ and $u_q = 0.77$ are then obtained. However, this result is meaningless, because the input data are *inconsistent*: the difference between the estimated values of Z_1 and Z_2 that, according to the model relations should be nearly zero, is not commensurate with their standard uncertainties. This may have come about because at least one of the reported estimated values failed to incorporate a systematic effect taking place during measurement or because the reported input uncertainties were not evaluated correctly, resulting in unreasonably low values. In this example we get $\chi^2_{min} = 6.557$ and $Bi = 2.561$. Multiplying the input uncertainties by the latter number produces again $q_e = 11.64$ but u_q increases from 0.77 to 1.97.

Example. Consider 'Peelle's puzzle' [5]‡. The measurand is modelled as $Q = X/Y$. Two independent measurement results for X are available, namely $x_{e1} = 1.00$ and $x_{e2} = 1.50$ with standard uncertainties $u_{x1} = 0.10$ and $u_{x2} = 0.15$ (arbitrary units). The measurement of Y gives $y_e = 1.0$ and $u_y = 0.2$.

The measurand can be evaluated in two different ways. Evaluator A obtains first an adjusted value for X using the model

$$\begin{pmatrix} X_1 - X \\ X_2 - X \end{pmatrix} = \mathbf{0}$$

which is of the form (5.31). The results are $x_{eA} = 1.154$ and $u_{xA} = 0.083$. Next he finds

$$q_{eA} = \frac{x_{eA}}{y_e} = 1.154$$

‡ This example follows a treatment by W Wöger.

and
$$u_{qA} = q_{eA}\left[\left(\frac{u_{xA}}{x_{eA}}\right)^2 + \left(\frac{u_y}{y_e}\right)^2\right]^{1/2} = 0.245.$$

Evaluator B uses the model
$$\begin{pmatrix} X_1/Y - Q \\ X_2/Y - Q \end{pmatrix} = \mathbf{0}$$

which is of the form (5.28) with $\mathbf{A} = (\,-1 \ -1\,)^T$ and input sensitivity matrix
$$\begin{pmatrix} 1/y_e & 0 & -x_{e1}/y_e^2 \\ 0 & 1/y_e & -x_{e2}/y_e^2 \end{pmatrix}$$

He/she then gets $q_{eB} = 0.882$ and $u_{qB} = 0.218$. Why the discrepancy? Who is right?

It is possible to gain insight on what is happening by deriving analytic expressions for each of the estimates of Q. After a bit of algebra we obtain
$$\frac{q_{eA} - q_{eB}}{q_{eA}} = \frac{\xi^2}{1+\xi^2}$$

where ξ is the ratio of the relative uncertainty of Y to the relative uncertainty of the difference $X_2 - X_1$, i.e.
$$\xi \equiv \frac{|x_{e1} - x_{e2}|}{(u_{x1}^2 + u_{x2}^2)^{1/2}} \frac{u_y}{|y_e|}.$$

It is thus clear that the estimate of evaluator B will always be less than that of A, the difference increasing as ξ grows. It is only for ξ very small that both estimates become similar. In this example, $\xi = 0.555$. If this value is decreased to 0.083, say by setting $u_y = 0.03$ (keeping all other input values constant), the results would be $q_{eA} = 1.154$ and $q_{eB} = 1.146$, with both evaluators reporting an uncertainty $u_q = 0.090$.

However, decreasing a stated uncertainty to make both results agree is quite an unnatural way to proceed. Besides, one should realize that these new results would still be meaningless, because the strong inconsistency in the given estimates for X has not been dealt with (the Birge ratio for the adjustment made by evaluator A turns out to be 2.774, whatever the value of u_y). If we wish to keep the estimates x_{e1} and x_{e2}, their uncertainties should be enlarged. Multiplying u_{x1} and u_{x2} by 2.774 and keeping the original data for y_e and u_y, evaluator A would obtain a new Birge ratio of 1 and the results $q_{eA} = 1.154$, $u_{qA} = 0.326$ while B would get $q_{eB} = 1.109$, $u_{qB} = 0.323$. A discrepancy still exists but it is much smaller than that obtained with the original inconsistent data.

It should finally be mentioned that, since the procedure of A involves no linearization, it may be claimed to be more accurate than that of B.

Of course, the consistency analysis applies to the results of the least-squares solution under uncertainty, when the input uncertainty matrix is part of the data. Sometimes, this matrix is not known, but it is known that all input quantities are uncorrelated and are measured with the same uncertainty σ, such that $\mathbf{u}_z^2 = \sigma^2 \mathbf{I}$. Then, the condition $\mathrm{Bi} = 1$ is used to find the 'reasonable' value

$$\sigma^2 = \frac{(\mathbf{z}_a - \mathbf{z}_e)^\mathrm{T}(\mathbf{z}_a - \mathbf{z}_e)}{n_\mathcal{M} - n_\mathcal{Q}}$$

and no further analysis is possible.

Example. Consider again the model relations $Q = Z_1$ and $Q = Z_2$ with $z_{e1} = 10$ and $z_{e2} = 14$. The input uncertainties are not given. Consistent least-squares adjustment is achieved using $u_{z1}^2 = u_{z2}^2 = \sigma^2 = 8$. Results are $q_e = 12$, and $u_q = 2$.

We conclude this section by noting that in curve-fitting problems, the algorithms presented in sections 5.6.1 and 5.6.2 are only approximate, because in both cases the solution is reached after a linearization procedure. Since the 'better' solution is the one that gives a lower value for $\chi^2_{\min}$, the Birge ratio can be taken as a figure of merit to judge their performances.

Example. Consider the measurement of the inductance λ and the internal resistance ρ of a black box that is connected in series with a known capacitor $\gamma = 0.02$ µF [42]. A sinusoidal voltage was applied to the combination at several frequency settings ω, which in this context is identified with the parameter τ in section 5.6. The phase shift θ of the capacitor voltage with respect to the applied voltage was read from an oscilloscope. With $X = \omega/\omega_o$ and $Y = \cot\theta$, ω_o being a reference frequency, the circuit's model is

$$Y = Q_1 X - Q_2 X^{-1}$$

where $Q_1 = \omega_o \lambda/\rho$ and $Q_2 = (\omega_o \rho \gamma)^{-1}$.

Input data corresponding to $\omega_o = 1$ rad s^{-1} are given in table 5.6. The estimates of Q_1 and Q_2, together with their standard and mutual uncertainties, are shown in table 5.7 (to avoid numerical instabilities, the programs were run with $\mathbf{x}'_e = \mathbf{x}_e/1000$). It is seen that, judging from the Birge ratio, in this example the method in section 5.6.2 gives a tighter but less consistent adjustment than that obtained with the method in section 5.6.1. Finally, table 5.8 shows the values and uncertainties of

148   The joint treatment of several measurands

Table 5.6. Input data for the electrical circuit.

i	x_{ei}	u_{xi}	y_{ei}	u_{yi}
1	22 000	440	−4.0170	0.50
2	22 930	470	−2.7420	0.25
3	23 880	500	−1.1478	0.08
4	25 130	530	1.4910	0.09
5	26 390	540	6.8730	1.90

Table 5.7. Parameters of the circuit's model.

Method	B_i	$q_{e1}/10^{-3}$	$q_{e2}/10^5$	$u_{q1}/10^{-3}$	$u_{q2}/10^5$	$r_{q1,q2}$
In section 5.6.1	0.854	1.0163	5.9372	0.1994	1.1161	0.9941
In section 5.6.2	0.843	1.0731	6.2499	0.2073	1.1606	0.9940

Table 5.8. Results for the electrical circuit.

Method	ρ_e/Ω	λ_e/mH	u_ρ/ρ_e	u_λ/λ_e	$r_{\rho,\lambda}$
In section 5.6.1	84.2	85.6	0.188	0.0224	−0.3168
In section 5.6.2	80.0	85.8	0.186	0.0220	−0.2877

the circuit's inductance and resistance. These are calculated from the models

$$\mathcal{M}_1(\rho, \lambda, q_1, q_2) = \rho - \frac{1}{q_2 \omega_o \gamma} = 0$$

and

$$\mathcal{M}_2(\rho, \lambda, q_1, q_2) = \lambda - \frac{q_1}{q_2 \omega_o^2 \gamma} = 0$$

together with (5.7) and sensitivity matrix

$$\mathbf{S} = \begin{pmatrix} 0 & \rho/q_2 \\ -\lambda/q_1 & \lambda/q_2 \end{pmatrix}.$$

It is seen that even though the estimates of Q_1 and Q_2 are strongly correlated, the correlation between the estimates of ρ and λ is much weaker.

5.8 Applications of least-squares adjustment

Four applications of least-squares adjustment are considered in this section. In the first, a method to evaluate the results of an intercomparison exercise is proposed. The second was discussed in [31] and deals with the calibration of a thermocouple. The third is taken from [35] and has to do (again) with comparison measurements. Finally, in the fourth application, taken from [34], the calibration of a mass balance is considered.

5.8.1 Reproducible measurements

Consider a *single* measurand Q that is measured under reproducible conditions. In 'round robin' intercomparisons, the measurand is an artifact that is circulated and measured in sequence by a series of participating laboratories. The objective of the exercise is not as much to determine the *reference value* q_e, but to see whether the results informed by the laboratories are comparable. By contrast, in the shared determination of fundamental constants, the objective is to find an *adopted value* for Q.

In these cases, the model relations are $Q = Z_i, i = 1, \ldots, n_Z$ or

$$\mathbf{1}Q = \mathbf{Z} \tag{5.41}$$

where $\mathbf{1}$ is vector of dimension $n_Z \times 1$ with all elements equal to one, n_Z is the number of laboratories and $\mathbf{z}_e$ consists of the estimates for the single quantity Q reported by each laboratory. Since (5.41) is of the form (5.31) with $\mathbf{A} = -\mathbf{1}$, we have

$$q_e = [\mathbf{1}^T(\mathbf{u}_z^2)^{-1}\mathbf{1}]^{-1}\mathbf{1}^T(\mathbf{u}_z^2)^{-1}\mathbf{z}_e$$
$$\mathbf{z}_a = \mathbf{1}q_e$$

and

$$u_q^2 = [\mathbf{1}^T(\mathbf{u}_z^2)^{-1}\mathbf{1}]^{-1}$$

where we recall that the right-hand side in the last expression means the summation of all the elements of $(\mathbf{u}_z^2)^{-1}$.

If, as is usually assumed, the input uncertainty matrix $\mathbf{u}_z^2$ is diagonal, the equations above reduce to

$$q_e = \left(\sum_{i=1}^{n_Z} \frac{1}{u_{zi}^2}\right)^{-1} \sum_{i=1}^{n_Z} \frac{z_{ei}}{u_{zi}^2} \tag{5.42}$$

$$u_q^2 = \left(\sum_{i=1}^{n_Z} \frac{1}{u_{zi}^2}\right)^{-1} \tag{5.43}$$

and if all individual uncertainties u_{zi} are equal to a (known) common value σ, (5.42) and (5.43) reduce to

$$q_e = \frac{1}{n}\sum_{i=1}^{n_z} z_{ei}$$

and

$$u_q^2 = \frac{\sigma^2}{n}.$$

Finally, if σ^2 is not known, consistent least-squares adjustment giving Bi $=$ 1 is achieved by choosing

$$\sigma^2 = \frac{1}{n-1}\sum_{i=1}^{n_z}(z_{ei} - q_e)^2.$$

Note that these results are exactly equal to those obtained with the TA analysis. However, they were derived here in a completely different way.

Example. Consider 10 independent measurements of the gravitational constant G [54]. The laboratories report the estimated values g_{ei} and standard uncertainties u_{gi} shown in table 5.9. The least-squares adjusted value is $g_e = 6.681\,76 \times 10^{-11}$ m^3 kg^{-1} s^{-2} with a standard uncertainty $u_g = 0.000\,26 \times 10^{-11}$ m^3 kg^{-1} s^{-2}. However, the unacceptably large Birge ratio 22.6 is due to the last value in table 5.9 being an outlier; it cannot be discarded since it corresponds to a measurement carried out very carefully and with modern means. Multiplication of the input uncertainties by Bi does not modify the best estimate but u_g increases to $0.005\,78 \times 10^{-11}$ m^3 kg^{-1} s^{-2}.

5.8.2 Calibration of a thermocouple

A thermocouple is an instrument that consists of two dissimilar metallic wires joined at one of their ends. This is called the measuring junction, also known as the 'hot' junction. The other ends are connected to a system maintained at a reference temperature T_r (normally the ice point temperature $0\,°C$, see figure 5.3). An electromotive force (emf) E is established across the wires when the temperature T at the measuring junction is different from the reference temperature.

The emf E, known as the 'Seebeck voltage', is a function of $T - T_r$ and depends on the material of the wires. To obtain this function, one measures the emfs E_i generated when the measurement junction is sequentially immersed in isothermal regions at temperatures T_i ($i = 1, \ldots, I$) while the reference junction is maintained at the reference temperature. The temperatures T_i may

Applications of least-squares adjustment 151

Table 5.9. Estimated values for the gravitational constant G in units 10^{-11} m^3 kg^{-1} s^{-2}.

i	g_{ei}	u_{gi}
1	6.672 59	0.000 85
2	6.674 0	0.000 7
3	6.672 9	0.000 5
4	6.687 3	0.009 4
5	6.669 9	0.000 7
6	6.674 2	0.000 7
7	6.683	0.011
8	6.675 4	0.001 5
9	6.673 5	0.002 9
10	6.715 40	0.000 56

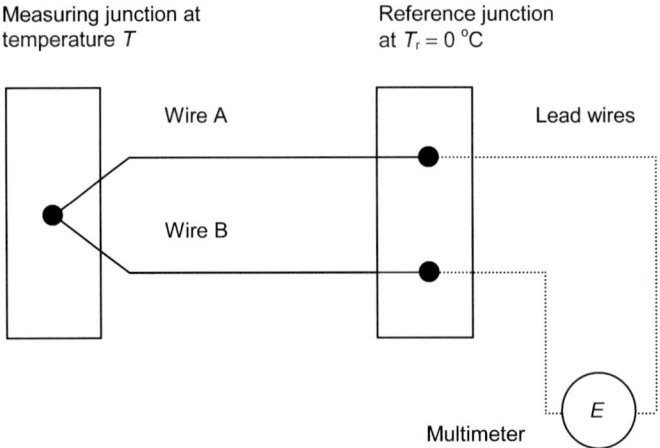

Figure 5.3. Schematic representation of a thermocouple.

be estimated from the readings of a calibrated thermometer, or by realizing a set of the thermometric fixed points of the International Temperature Scale of 1990 (ITS-90) [44]. These data are then used to construct the function $E(T)$ such that the values $E(T_i)$ are very nearly equal to the corresponding measured values E_i, corrected for systematic effects (figure 5.4).

The form of the calibration curve $E(T)$ is not standardized. Calibration laboratories customarily take

$$E = \sum_{j=1}^{J} Q_j T^j \tag{5.44}$$

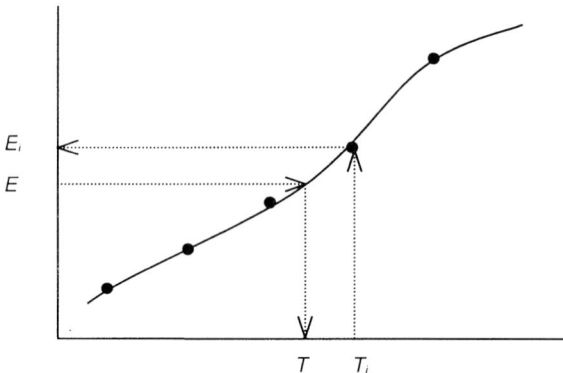

Figure 5.4. Calibration curve of a thermocouple.

where J is at most equal to the number I of calibration points and T is in degrees Celsius. This forces the emf at $0\,°C$ to be zero. Normally, a cubic or quartic polynomial is chosen as the calibration curve.

To find the values of $\mathbf{Q} = (Q_1 \ \ldots \ Q_J)^T$ we use the methods in section 5.6 with $\mathbf{X} = \mathbf{T} = (T_1 \ \ldots \ T_I)^T$, $\mathbf{Y} = \mathbf{E} = (E_1 \ \ldots \ E_I)^T$ and

$$\mathbf{F(T)} = \begin{pmatrix} T_1 & \cdots & T_1^J \\ \vdots & \ddots & \vdots \\ T_I & \cdots & T_I^J \end{pmatrix}.$$

Example. Consider the calibration of a type S thermocouple. $I = 11$ calibration temperatures at $18\,°C$ intervals were selected. These are shown in table 5.10, together with the corresponding emf values and their respective uncertainties. A fourth degree polynomial of the form (5.44) was used for the adjustment, the results of which are shown in table 5.11, where the units of q_{ej} are $\mu V/(°C)^j$. Note the substantial increase of the relative uncertainty associated with the estimates of the higher order coefficients. In this case, both algorithms in section 5.6 yielded very similar results, with $Bi = 1.08$.

In practice, the calibration laboratory should report not only the estimated values of the coefficients of the calibration curve and their associated uncertainty matrix, but also the calibration curve itself, both graphically and in the form of a table to facilitate interpolation. This is because when the thermocouple is in service, the user measures an emf value E, to which a temperature T corresponds according to equation (5.44), which is not explicit.

To evaluate the standard uncertainty associated with the estimates of temperature obtained by the user, the following procedure, taken from [31], is

Applications of least-squares adjustment 153

Table 5.10. Calibration points and measured emfs.

Point	$T_i/°C$	$u(T_i)/°C$	$E_i/\mu V$	$u(E_i)/\mu V$
1	0	0.04	0	1.20
2	18	0.04	108.1	1.20
3	36	0.04	222.1	1.20
4	54	0.08	338.7	1.20
5	72	0.08	462.9	1.20
6	90	0.08	587.2	1.38
7	108	0.12	720.6	1.38
8	126	0.12	851.7	1.38
9	144	0.12	993.6	1.38
10	162	0.20	1130.8	1.56
11	180	0.20	1280.3	1.56

Table 5.11. Coefficients of the fourth-order calibration curve of the thermocouple.

j	q_{ej}	u_{rj}	Correlation coefficients		
1	5.8631	1.3 %	−0.967	0.914	−0.862
2	8.574×10^{-3}	28.3 %		−0.986	0.956
3	-1.44×10^{-5}	160.0 %			−0.992
4	2.9×10^{-8}	234.4 %			

recommended. Consider E, T and $\mathbf{q}_e$ as independent variables in

$$f(E, T, \mathbf{q}_e) = E - g(T, \mathbf{q}_e) = 0 \tag{5.45}$$

where

$$g(T, \mathbf{q}_e) = \sum_{j=1}^{J} q_{ej} T^j$$

and E is the emf generated when the thermocouple is in service. Now, (5.45) can be considered as defining T as an implicit function of E and $\mathbf{q}_e$. If T was solved from that function, it would be written as $T = h(E, \mathbf{q}_e)$. It follows from the LPU, equation (3.22), that

$$u^2(T) = u_u^2 + u_c^2$$

where

$$u_u^2 = \left(\frac{\partial h}{\partial E}\right)^2 u^2(E)$$

is the square of what may be called the 'user's uncertainty' (since it depends on data gathered while the thermocouple is in actual use) and

$$u_c^2 = \mathbf{h}_q^T \mathbf{u}_q^2 \mathbf{h}_q \tag{5.46}$$

is the square of the 'calibration uncertainty', so-called because it depends only on the data reported in the calibration certificate.

In (5.46), the components of the $J \times 1$ vector $\mathbf{h}_q$ are the partial derivatives $\partial h/\partial q_j$ evaluated at $\mathbf{q} = \mathbf{q}_e$. Now of course, these derivatives cannot be evaluated directly, because the function $h(E, \mathbf{q})$ is implicit. We then resort to the standard rules of calculus (see e.g. [19, p 347]), from which

$$\frac{\partial h}{\partial E} = -\frac{\partial f/\partial E}{\partial f/\partial T} = -\frac{1}{\partial g/\partial T} = -\frac{1}{S}$$

where

$$S = \frac{\partial g}{\partial T} = \frac{\partial E}{\partial T}$$

is called the *Seebeck coefficient*. Similarly,

$$\frac{\partial h}{\partial q_j} = -\frac{\partial f/\partial q_j}{\partial f/\partial T} = -\frac{\partial g/\partial q_j}{\partial g/\partial T} = -\frac{T^j}{S}.$$

From this last expression it may be seen that u_c will always be zero at $T = 0\,°C$. This is because (5.44) constrains the calibration function to vanish at the ice point temperature. Nevertheless, the contribution of u_u to $u(T)$ will result in a total uncertainty that is different from zero at all temperatures.

In summary, then,

$$u^2(T) = \frac{1}{S^2}[u^2(E) + \mathbf{T}^T \mathbf{u}_q^2 \mathbf{T}] \tag{5.47}$$

where

$$\mathbf{T} = (T \quad T^2 \quad \ldots \quad T^J)^T.$$

In order for the user to evaluate (5.47) the value of the Seebeck coefficient is needed; it can be computed from

$$S = \sum_{j=1}^{J} j q_{ej} T^{j-1}$$

although this effort can be saved if S is given (in the form of a table) in the calibration certificate.

Example. Assume that with the thermocouple calibrated as before the user measures a value $E = 610.8\,\mu V$ with $u(E) = 2.1\,\mu V$. The temperature corresponding to this emf is found from the calibration curve supplied in the calibration certificate, and turns out to be $T = 93.11\,°C$. The corresponding Seebeck coefficient is $S = 7.1787\,\mu V/°C$. Then, $u_u = 0.2925\,°C$, $u_c = 0.1176\,°C$ and $u(T) = 0.32\,°C$.

5.8.3 Difference comparison measurements revisited

Difference comparison measurements have already appeared in section 3.11.1 and in section 5.4. In these, an attribute T of a test specimen (for example, its mass) and the corresponding attribute S of a reference standard are measured one after the other. The resulting values t and s are not necessarily the best estimates of T and S, because the comparator readings may represent differences with an unknown but stable quantity (for example, a tare weight) and are generally affected by various factors. Nonetheless, in calculating the value of the difference $D = T - S$ these constant influences are supposedly cancelled. The conventional true value t_t of T is estimated as $t_t = s_t + d$, where s_t is the known conventional true value of the standard and $d = t - s$.

The value of D is usually estimated from redundant measurements of the quantities S and T using sequences such as $(STS)_n$, $(TST)_n$, $(STTS)_n$, $(TSST)_n$, and so forth, where n is a repeat factor for the cycle in parentheses. Moreover, the reference standard S can be used to measure comparison differences $D_j = T_j - S$ for several test specimens T_j, $j = 1, \ldots, m$, again using different cyclic combinations such as $(ST_1 \ldots T_m S)_n$, $(ST_1 ST_2 \ldots T_m S)_n$, etc. It is also possible to involve more than one measurement standard in the comparison sequence.

In the following discussion we present two alternative procedures to analyse this problem. An illustrative example then follows.

Method 1

In this method, taken from [48], we let $\mathbf{Q}$ be the vector of n_Q quantities involved in the comparison. The measurement sequence is defined by the transformation

$$\mathbf{P} = \mathbf{LQ}$$

where $\mathbf{P}$ is a vector of n_P time-ordered quantities and $\mathbf{L}$ is the $n_P \times n_Q$ comparison design matrix. All elements of this matrix are null except the one element equal to 1 in each line that sets the measurement order. Therefore, a quantity Q_i corresponds to several quantities P_k and a given difference $Q_i - Q_j$ can be evaluated from various pairs of differences $P_k - P_l$.

Define now the vectors $\mathbf{G}$ and $\mathbf{C}$ consisting of the $n_G = n_C = n_P$ quantities G_i and C_i that represent, respectively, the information rendered by the comparator and the corrections for systematic effects occurring when P_i is measured. The measurement model is thus

$$\mathbf{P} = \mathbf{G} + \mathbf{C}.$$

As was mentioned earlier, the advantage of comparison measurements is that constant systematic effects common to all measurements need not be modelled, because their influence on the determination of the differences cancels out. However, it may be necessary to apply particular corrections to each reading (for example, buoyancy corrections in mass measurements). Furthermore, the

instrument may be affected by drift, i.e. time-dependent systematic effects. It is assumed here that all necessary time-constant corrections are included in the estimates of **G**, and that **C** stands for temporal drift corrections only. The latter may be modelled as
$$C_i = F_1\theta_i + \cdots + F_{n_F}(\theta_i)^{n_F}$$
where the F_js are unknown coefficients and θ_i is the time interval between the measurements of P_i and P_1 (it must be assumed that these intervals are perfectly known, otherwise they become further input quantities to the measurement model and the problem gets nonlinear). The order n_F of the drift polynomial can be chosen at will, although it would seem unreasonable to go beyond $n_F = 3$ and in any case, because of (5.27), it must be $n_F < n_G - n_Q$. Equivalently,
$$\mathbf{C} = \mathbf{\Theta F}$$
where $\mathbf{\Theta}$ is the $n_G \times n_F$ matrix
$$\mathbf{\Theta} = \begin{pmatrix} 0 & \cdots & 0 \\ \theta_2 & \cdots & (\theta_2)^{n_F} \\ \vdots & \ddots & \vdots \\ \theta_{n_G} & \cdots & (\theta_{n_G})^{n_F} \end{pmatrix}$$
and $\mathbf{F} = (F_1 \quad \cdots \quad F_{n_F})^T$.

The measurement model becomes
$$\mathbf{LQ} = \mathbf{G} + \mathbf{\Theta F}$$
or
$$\mathbf{AO} = \mathbf{G}$$
where **O** and **A** are the block matrices $(\mathbf{Q} \quad \mathbf{F})^T$ and $(\mathbf{L} \quad -\mathbf{\Theta})$, respectively.

Since this model is of the form (5.31), use of the pertinent equations yields
$$\mathbf{o}_e = [\mathbf{A}^T(\mathbf{u_g^2})^{-1}\mathbf{A}]^{-1}\mathbf{A}^T(\mathbf{u_g^2})^{-1}\mathbf{g}_e \tag{5.48}$$
$$\mathbf{u_o^2} = [\mathbf{A}^T(\mathbf{u_g^2})^{-1}\mathbf{A}]^{-1} \tag{5.49}$$
where $\mathbf{g}_e$ is the vector of input measured values (corrected for constant systematic effects), $\mathbf{o}_e$ are the estimates of **O** and $\mathbf{u_g^2}$ and $\mathbf{u_o^2}$ are the corresponding uncertainty matrices.

Finally, let **D** be the vector of n_D differences of interest $Q_j - Q_k$, where usually Q_j is a test quantity and Q_k is a reference quantity. The relation between **D** and **Q** is
$$\mathbf{D} = \mathbf{HQ}$$
where **H** is an $n_D \times n_Q$ matrix whose lines contain one element equal to -1, one element equal to 1 and the rest of the elements are null. Recovering the estimates $\mathbf{q}_e$ and the uncertainty matrix $\mathbf{u_q^2}$ from $\mathbf{o}_e$ and $\mathbf{u_o^2}$, respectively, we have
$$\mathbf{d}_e = \mathbf{Hq}_e$$
$$\mathbf{u_d^2} = \mathbf{Hu_q^2H}^T.$$

Method 2

An alternative method consists of formulating the $n_\mathcal{M}$ model relations that give directly the n_D differences $\mathbf{D}$ in terms of the n_G input quantities $\mathbf{G}$, with $n_D < n_\mathcal{M} \leq n_G$, see (5.27). This model is

$$\mathbf{ED} = \mathbf{BG} \qquad (5.50)$$

where $\mathbf{E}$ is an $n_\mathcal{M} \times n_D$ order matrix and $\mathbf{B}$ is an $n_\mathcal{M} \times n_G$ evaluation matrix. The structure of these matrices depends on the type and number of cycles n and on the way the indications are processed, see the numerical example later.

The model (5.50) is of the form (5.29). Then

$$\mathbf{d}_e = (\mathbf{E}^T \mathbf{K} \mathbf{E})^{-1} \mathbf{E}^T \mathbf{K} \mathbf{B} \mathbf{g}_e \qquad (5.51)$$
$$\mathbf{u}_d^2 = (\mathbf{E}^T \mathbf{K} \mathbf{E})^{-1} \qquad (5.52)$$

where $\mathbf{K} = (\mathbf{B} \mathbf{u}_g^2 \mathbf{B}^T)^{-1}$.

This second method does not yield estimates for the individual quantities, nor does it take drift explicitly into account. However, if the measurements are taken equally spaced in time, both the linear and quadratic components of drift may be eliminated by a judicious choice of matrix $\mathbf{B}$.

Equal variance measurements

The input data to both methods consist of the values $\mathbf{g}_e$ and the uncertainty matrix $\mathbf{u}_g^2$. No general rules can be given here with regards to the latter, since its structure depends on the particulars of the comparison (for example, buoyancy corrections in mass measurements will introduce non-zero off-diagonal terms to $\mathbf{u}_g^2$). However, if all individual measurements are independent the input uncertainty matrix will be diagonal. If it is deemed reasonable to assume further that the uncertainty associated with all values $\mathbf{g}_e$ is equal to a common value σ, $\mathbf{u}_g^2$ becomes equal to σ^2 times the $n_I \times n_I$ unit matrix. In this case (5.48) and (5.49) reduce to

$$\mathbf{o}_e = (\mathbf{A}^T \mathbf{A})^{-1} \mathbf{A}^T \mathbf{g}_e$$
$$\mathbf{u}_o^2 = \sigma^2 (\mathbf{A}^T \mathbf{A})^{-1}$$

while (5.51) and (5.52) become

$$\mathbf{d}_e = (\mathbf{E}^T \mathbf{K}' \mathbf{E})^{-1} \mathbf{E}^T \mathbf{K}' \mathbf{B} \mathbf{g}_e$$
$$\mathbf{u}_d^2 = \sigma^2 (\mathbf{E}^T \mathbf{K}' \mathbf{E})^{-1}$$

where $\mathbf{K}' = (\mathbf{B}\mathbf{B}^T)^{-1}$. Therefore, the common estimate of variance σ^2 will have no effect on the output estimates, acting only as a multiplicative factor in the output uncertainty matrix. In accordance with the Birge ratio criterion, σ^2 should not deviate much from

$$\sigma_1^2 = \frac{(\mathbf{A}\mathbf{o}_e - \mathbf{g}_e)^T(\mathbf{A}\mathbf{o}_e - \mathbf{g}_e)}{n_G - n_O}$$

in method 1 or from

$$\sigma_2^2 = \frac{(\mathbf{E}\mathbf{d}_e - \mathbf{B}\mathbf{g}_e)^T \mathbf{K}'(\mathbf{E}\mathbf{d}_e - \mathbf{B}\mathbf{g}_e)}{n_M - n_D}$$

in method 2.

The variance estimate σ^2 should include at least both the repeatability of the comparator readings and its resolution. The repeatability component has to be based on repeated measurements of *one* of the quantities S or T, and if significant time-varying systematic effects are suspect these measurements should be carried out after resetting each time the comparator to its initial conditions. Alternatively one can force consistency by taking σ^2 close or equal to σ_1^2 or σ_2^2 (according to the chosen evaluation procedure). In that case it should be checked whether the resulting uncertainty is commensurable with the repeatability and resolution characteristics of the comparator.

Example. Table 5.12, taken from [48], gives the comparator readings obtained from the measurement of four 1 kg masses using the sequence $(ST_1T_2T_3)_3$. It is assumed here that no corrections for systematic effects other than drift are necessary and that S is the only reference quantity. The readings were taken at equal time intervals $\theta_i/(\text{t.u.}) = i - 1$, where t.u. stands for 'time units'.

In method 1 we write

$$\mathbf{Q} = (S \quad T_1 \quad T_2 \quad T_3)^T$$

and $\mathbf{L}$ is obtained by stacking three 4×4 identity matrices. Then $n_P = 12$ and

$$\mathbf{P} = (S \quad T_1 \quad T_2 \quad T_3 \quad S \quad T_1 \quad \ldots \quad T_3)^T.$$

Further, we consider only the linear and quadratic components of drift, thus $n_F = 2$. We then have

$$\Theta = \begin{pmatrix} 0 & 1 & 2 & \ldots \\ 0 & 1 & 4 & \ldots \end{pmatrix}^T$$

and

$$\mathbf{H} = \begin{pmatrix} -1 & 1 & 0 & 0 \\ -1 & 0 & 1 & 0 \\ -1 & 0 & 0 & 1 \end{pmatrix}.$$

In method 2 we use

$$\mathbf{D} = \begin{pmatrix} T_1 - S \\ T_2 - S \\ T_3 - S \end{pmatrix}$$

and take $n_{\mathcal{M}} = 6$ model relations with

$$\mathbf{E} = \begin{pmatrix} 1 & 1 & 0 & 0 & 0 & 0 \\ 0 & 0 & 1 & 1 & 0 & 0 \\ 0 & 0 & 0 & 0 & 1 & 1 \end{pmatrix}^{\mathrm{T}}$$

and

$$\mathbf{B} = \begin{pmatrix} a & b & 0 & 0 & c & d & 0 & 0 & 0 & 0 & 0 \\ 0 & 0 & 0 & 0 & a & b & 0 & 0 & c & d & 0 & 0 \\ e & 0 & f & 0 & g & 0 & h & 0 & 0 & 0 & 0 & 0 \\ 0 & 0 & 0 & 0 & e & 0 & f & 0 & g & 0 & h & 0 \\ i & 0 & 0 & j & k & 0 & 0 & l & 0 & 0 & 0 & 0 \\ 0 & 0 & 0 & 0 & i & 0 & 0 & j & k & 0 & 0 & l \end{pmatrix}$$

where $a, b, \ldots$ are weight factors satisfying $a + c = -1$, $b + d = 1$, etc. This gives

$$T_1 - S = (bG_2 + dG_6) + (aG_1 + cG_5)$$
$$T_1 - S = (bG_6 + dG_{10}) + (aG_5 + cG_9)$$

with G_1, G_5 and G_9 corresponding to S, and G_2, G_6 and G_{10} corresponding to T_1. Similar equations express the differences $T_2 - S$ and $T_3 - S$.

It is easily verified that, in this example, the linear and quadratic components of drift are eliminated by choosing

$$a = -d = -\tfrac{3}{8} \qquad b = -c = \tfrac{5}{8}$$
$$e = -h = -\tfrac{1}{4} \qquad f = -g = \tfrac{3}{4}$$
$$i = -l = -\tfrac{1}{8} \qquad j = -k = \tfrac{7}{8}.$$

Table 5.13 gives the estimates and uncertainty matrix for the comparison quantities obtained from method 1. These values estimate the difference between each of the masses and the unknown but stable tare mass of the comparator. Table 5.14 shows the estimates of the differences of masses T_1 through T_3 with respect to mass S. Results were obtained using $\sigma = 0.39$ µg. This value appears to be reasonable because it is between $\sigma_1 = 0.42$ µg and $\sigma_2 = 0.36$ µg and is of the order of the resolution of the comparator.

Table 5.12. Indicated values.

Cycle	P	$g_e/\mu g$
1	$P_1 = S$	22.1
	$P_2 = T_1$	743.7
	$P_3 = T_2$	3080.4
	$P_4 = T_3$	4003.4
2	$P_5 = S$	18.3
	$P_6 = T_1$	739.2
	$P_7 = T_2$	3075.5
	$P_8 = T_3$	3998.2
3	$P_9 = S$	14.2
	$P_{10} = T_1$	734.7
	$P_{11} = T_2$	3071.6
	$P_{12} = T_3$	3994.8

Table 5.13. Comparison quantities (from method 1). S and the T_is are in µg, F_1 in µg/(t.u.) and F_2 in µg/(t.u.)², where t.u. means 'time units'.

				Correlation coefficients				
i	Quantity	Value	Uncertainty	$i = 2$	$i = 3$	$i = 4$	$i = 5$	$i = 6$
1	S	22.56	0.32	0.549	0.562	0.555	0.672	−0.576
2	T_1	744.64	0.35		0.606	0.602	0.704	−0.590
3	T_2	3082.35	0.37			0.632	0.696	−0.561
4	T_3	4006.38	0.38				0.650	−0.491
5	F_1	1.12	0.12					−0.960
6	F_2	−0.004	0.011					

5.8.4 Calibration of a balance

Balances used in trade and commerce are usually not calibrated, they are only adjusted and verified, as explained in section 3.11.3. For more exacting measurements, however, it may be necessary to establish the calibration curve of the instrument. This is done by means of a mass scale, which consists of a set of known weights that are placed on the balance. The differences between the recorded indications and the masses of the scale serve to construct the calibration curve.

The weights of the scale are, in turn, determined by comparison weighings using one or more reference masses of known conventional true value. Very often

Applications of least-squares adjustment

Table 5.14. Comparison differences.

Method	i	Difference	Value/µg	Uncert./µg	Corr. coeff. $i=2$	$i=3$
1	1	$T_1 - S$	722.08	0.32	0.511	0.505
	2	$T_2 - S$	3059.79	0.33		0.529
	3	$T_3 - S$	3983.82	0.33		
2	1	$T_1 - S$	721.99	0.34	0.493	0.484
	2	$T_2 - S$	3059.61	0.35		0.491
	3	$T_3 - S$	3983.55	0.36		

only one standard is used. In this case the nominal values of the scale must be in certain specific relations to the mass of the standard. For example, a 1 kg standard may be used to calibrate a five-mass scale consisting of the nominal values 100 g, 100 g, 200 g, 200 g and 500 g. The weighing scheme may be

$$m_{1\,\text{kg}} - (m_{100\,\text{g}} + m_{200\,\text{g}} + m'_{200\,\text{g}} + m_{500\,\text{g}}) = d_1$$
$$m_{500\,\text{g}} - (m'_{100\,\text{g}} + m_{200\,\text{g}} + m'_{200\,\text{g}}) = d_2$$
$$m_{200\,\text{g}} - m'_{200\,\text{g}} = d_3$$
$$m'_{200\,\text{g}} - (m_{100\,\text{g}} + m'_{100\,\text{g}}) = d_4$$
$$m_{100\,\text{g}} - m'_{100\,\text{g}} = d_5$$

where the d_is are the indications of the comparator.

This system can be solved readily, because it consists of five equations for the five unknowns. However, if two additional weighings

$$m_{200\,\text{g}} - (m_{100\,\text{g}} + m'_{100\,\text{g}}) = d_6$$
$$m_{500\,\text{g}} - (m_{100\,\text{g}} + m_{200\,\text{g}} + m'_{200\,\text{g}}) = d_7$$

are carried out, the system is overdetermined and a least-squares adjustment becomes necessary.

This procedure yields values that are strongly correlated. Accordingly, the mutual uncertainties must be taken into account when the mass scale is used to construct the calibration curve.

The following examples illustrate (i) the procedure to determine a seven-mass scale, (ii) the calibration of a weighing instrument using the scale and (iii) the use of the resulting calibration curve.

Example. A reference standard of mass $m_r = 1\,\text{kg} + 1.10\,\text{mg}$ and standard uncertainty $u_{mr} = 0.02\,\text{mg}$ was used to calibrate a set of

162   The joint treatment of several measurands

Table 5.15. Nominal values of the mass scale.

Mass	Nominal value
1	1000 g
2	500 g
3	500 g
4	200 g
5	200 g
6	100 g
7	100 g

seven masses $\mathbf{m}$, the nominal values of which are shown in table 5.15. Comparison measurements were carried out with the scheme $\mathbf{L_m m'}$, where $\mathbf{m'} = (m_r \quad \mathbf{m}^T)^T$ and

$$\mathbf{L_m} = \begin{pmatrix} 1 & -1 & 0 & 0 & 0 & 0 & 0 & 0 \\ 1 & 0 & -1 & -1 & 0 & 0 & 0 & 0 \\ 0 & 1 & -1 & -1 & 0 & 0 & 0 & 0 \\ 0 & 0 & 1 & -1 & 0 & 0 & 0 & 0 \\ 0 & 0 & 1 & 0 & -1 & -1 & -1 & 0 \\ 0 & 0 & 0 & 1 & -1 & -1 & 0 & -1 \\ 0 & 0 & 0 & 0 & 1 & -1 & 0 & 0 \\ 0 & 0 & 0 & 0 & 1 & 0 & -1 & -1 \\ 0 & 0 & 0 & 0 & 0 & 1 & -1 & -1 \\ 0 & 0 & 0 & 0 & 0 & 0 & 1 & -1 \end{pmatrix}$$

is the matrix that specifies the loading scheme. The estimated mass differences $\mathbf{d_e}$ and associated standard uncertainties are shown in table 5.16. (The weighings were assumed to be uncorrelated.)

Instead of determining the values $\mathbf{m}$ directly, it is better to obtain the quantities $\mathbf{Q} = \mathbf{m} - \mathbf{m_n}$, where $\mathbf{m_n}$ is the vector of nominal values in table 5.15. This is done by letting $\Delta m_r = 1.1$ mg and defining $\mathbf{A}$ as the 10×7 matrix formed by eliminating the first column of $\mathbf{L_m}$, $\mathbf{Z}$ as the vector with estimates

$$z_{ei} = \begin{cases} d_{ei} - \Delta m_r & \text{for } i = 1, 2 \\ d_{ei} & \text{for } i = 3, \ldots, 10 \end{cases}$$

and $\mathbf{u}_z^2$ as the diagonal input uncertainty matrix with components

$$u_{zi}^2 = \begin{cases} u_{di}^2 + u_{mr}^2 & \text{for } i = 1, 2 \\ u_{di}^2 & \text{for } i = 3, \ldots, 10. \end{cases}$$

Applications of least-squares adjustment

Table 5.16. Observed mass differences $\mathbf{d_e}$.

i	d_{ei}/mg	u_{di}/mg
1	−0.2031	0.0032
2	−0.1984	0.0035
3	−0.0035	0.0026
4	0.0972	0.0030
5	−0.0061	0.0035
6	−0.0455	0.0029
7	0.0495	0.0018
8	0.0006	0.0019
9	−0.0509	0.0015
10	0.0496	0.0010

Table 5.17. Mass scale results $\mathbf{q_e}$ and components of the uncertainty matrix $\mathbf{u_q^2}$.

			\multicolumn{6}{c}{Correlation coefficients}					
i	q_{ei}/mg	u_{qi}/mg	$i=2$	$i=3$	$i=4$	$i=5$	$i=6$	$i=7$
1	1.2990	0.0144	0.9691	0.9691	0.9084	0.9130	0.8533	0.8573
2	0.6986	0.0073		0.9405	0.9046	0.9091	0.8558	0.8475
3	0.6038	0.0073			0.9145	0.9190	0.8530	0.8690
4	0.2907	0.0031				0.8922	0.8501	0.8491
5	0.2404	0.0031					0.8731	0.8722
6	0.1704	0.0017						0.8237
7	0.1205	0.0017						

The model relations are then of the form (5.31)

$$\mathbf{AQ} - \mathbf{Z} = \mathbf{0}.$$

A Birge ratio equal to 0.95 is obtained for the adjustment, the results of which appear in table 5.17.

Example. The mass scale in the previous example was used to calibrate a weighing instrument at seven points in the range 0–2 kg at 300 g intervals. Define $\mathbf{Y}$ as the vector of masses placed on the instrument and $\mathbf{X}$ as the corresponding vector of indications. With this notation,

Table 5.18. Observed indications x_e.

i	x_{ei}/g	u_{xi}/g
1	1996.626	0.016
2	1697.955	0.014
3	1398.891	0.010
4	1099.463	0.009
5	799.799	0.007
6	499.934	0.005
7	199.990	0.004

the calibration scheme is written as $\mathbf{Y} = \mathbf{L}_c \mathbf{m}$, where

$$\mathbf{L}_c = \begin{pmatrix} 1 & 0 & 1 & 1 & 1 & 0 & 1 \\ 1 & 1 & 0 & 0 & 1 & 0 & 0 \\ 0 & 1 & 1 & 1 & 0 & 1 & 1 \\ 0 & 0 & 1 & 1 & 1 & 1 & 1 \\ 0 & 1 & 0 & 0 & 1 & 0 & 1 \\ 0 & 0 & 0 & 1 & 1 & 1 & 0 \\ 0 & 0 & 0 & 1 & 0 & 0 & 0 \end{pmatrix}.$$

The data for $\mathbf{X}$ are shown in table 5.18. The data for $\mathbf{Y}$ are obtained as $\mathbf{y}_e = \mathbf{L}_c(\mathbf{q}_e + \mathbf{m}_n)$. Since the nominal values carry no uncertainty, $\mathbf{u}_y^2 = \mathbf{L}_c \mathbf{u}_q^2 \mathbf{L}_c^T$. Note that $\mathbf{L}_c$ cannot have more than seven lines, otherwise $\mathbf{u}_y^2$ would be singular.

For the calibration curve of the instrument a homogeneous polynomial of degree $J \leq 7$ was selected, i.e.

$$Y = Q_1 X + Q_2 X^2 + \cdots + Q_J X^J$$

where Y is a quantity that represents the mass submitted to the instrument *in use*, X is the corresponding indicated value ($x < 2$ kg) and the Q_is are the coefficients to be determined.

Adjustments with J varying from 1 through 5 were made (to avoid numerical instabilities, the programs were run with mass values in kg). The resulting Birge ratios are shown in table 5.19. These values indicate that in this case the method in section 5.6.1 gives a 'better' adjustment, and that a cubic polynomial is an adequate choice for the calibration curve. Table 5.20 shows the results of the adjustment using both methods for $J = 3$.

Example. In use, the instrument calibrated as before indicates the value $x_e = 475.892$ g with a standard uncertainty $u_x = 0.030$ g. This

Applications of least-squares adjustment

Table 5.19. Birge ratios.

J	By method in section 5.6.1	Vy method in section 5.6.2
1	65.7	127
2	17.1	41.5
3	1.0	1.2
4	1.1	1.4
5	1.2	1.3

Table 5.20. Coefficients $\mathbf{q}_e$ and components of the uncertainty matrix $\mathbf{u}_\mathbf{q}^2$ for the cubic calibration curve.

				Corr. coeff.	
Method	j	q_{ej}	u_{qj}	$j=2$	$j=3$
in section 5.6.1	1	1.000 0893	0.000 0183	-0.9447	0.8694
	2	$-0.000\,1822$ kg^{-1}	0.000 0322 kg^{-1}		-0.9790
	3	0.000 4931 kg^{-2}	0.000 0129 kg^{-2}		
in section 5.6.2	1	1.000 0660	0.000 0183	-0.9447	0.8694
	2	$-0.000\,1438$ kg^{-1}	0.000 0322 kg^{-1}		-0.9790
	3	0.000 4792 kg^{-2}	0.000 0129 kg^{-2}		

corresponds to a mass $y_e = \mathbf{F}(x_e)^T \mathbf{q}_e$, where $\mathbf{F}(x_e)$ is the vector

$$\mathbf{F}(x_e) = (\, x_e \quad x_e^2 \quad \ldots \quad x_e^J \,)^T.$$

The associated standard uncertainty is obtained as

$$u_y^2 = u_u^2 + u_c^2$$

where

$$u_u^2 = (q_{e1} + 2q_{e2}x_e + \cdots + Jq_{eJ}x_e^{J-1})^2 u_x^2$$

is a component of uncertainty attributable to the user's measurement system and

$$u_c^2 = \mathbf{F}(x_e)^T \mathbf{u}_\mathbf{q}^2 \mathbf{F}(x_e)$$

is a component of uncertainty attributable to the calibration procedure. With $J = 3$, the results are $y_e = 475.946$ g with the method in section 5.6.1 and $y_e = 475.942$ g with the method in section 5.6.2. In both cases $u_y = 0.030$ g.

Chapter 6

Bayesian inference

Both the standard uncertainty and the expanded uncertainty associated with the best estimate of a measurand depend on the probability density function, pdf, that describes the state of knowledge about the quantity. Up to this point, the normal and the rectangular pdfs have been used on rather intuitive grounds. Other 'special' pdfs, like the log pdf in section 3.3.3 and the 'U' pdf in section 3.5.3, have been derived without much effort. However, obtaining the most appropriate pdf for a general application is not normally an easy task.

Fortunately, in most cases the pdf is not a *sine qua non* requirement. The reason is that we may usually obtain the standard uncertainty through the application of the LPU, while the expanded uncertainty, if needed, may be evaluated using the *ad hoc* procedure in chapter 4. However, the LPU cannot be applied to primary quantities that are directly measured. Even though in such cases we are allowed to use the TA evaluation procedure, the number of individual measurements must be 'high'. Moreover, when TA-evaluated uncertainties are combined with those evaluated 'by other methods' (type B), the theoretical meaning and the practical use of the resulting uncertainty are unclear, especially if a coverage interval is required. Furthermore, the treatment in previous chapters does not allow us to incorporate into the evaluation procedure any prior knowledge about the measurand that we may have. This drawback is very serious, because although we often do not start by being completely ignorant about the quantity or quantities of interest (for example, in recalibration), the conventional procedure forces us to throw away this prior information.

In this chapter we shall present a general method to obtain the pdf that most adequately expresses what we know about the measurand, both before and after measurement. The method is based on the theorem of Bayes; it allows us to refine our current state of knowledge about the measurand in the light of the new information acquired through measurement experiments.

Bayes' theorem is introduced in sections 6.1 and 6.2; two simple illustrative examples are given in section 6.3. For those cases when a state of complete ignorance is our starting point, non-informative priors are introduced in section 6.4. Section 6.5 presents different alternatives to evaluate the best estimate and the standard uncertainty associated with the estimate of a quantity that is repeatedly measured. The alternative to use depends on the assumed underlying frequency distribution from which the measurement values are presumed to have been taken. Then, in section 6.6, a summary explanation of the principle of maximum entropy is provided. This principle gives us a criterion to select, among all pdfs that seemingly satisfy a given set of restrictions, the one that does not provide more information than we have about the measurand. Formulae for the consumer's and producer's risks are next derived in section 6.7. Finally, in section 6.8, the concept of a 'model prior' is introduced. This is applied, in section 6.9, to the Bayesian evaluation of the common model $Z = \mathcal{M}(X, Y)$.

6.1 Bayes' theorem (BT)

Consider the following axiom of probability theory: given two propositions A and B, the probability of both of them being true is equal to the product of the probability of A being true multiplied by the probability of B being true given that A is true, and *vice versa* (figure 6.1(*a*)). In symbols:

$$p(AB) = p(A)p(B|A) = p(B)p(A|B)$$

and therefore

$$p(B|A) = \frac{p(B)p(A|B)}{p(A)} \qquad (6.1)$$

where '|' means 'conditional upon' or 'given' for short.

Result (6.1) is known as *Bayes' theorem* (BT). It is usually covered in introductory chapters of books on sampling theory (e.g. [38]), where it is typically presented by taking B as one of a set of mutually exclusive propositions B_i, $i = 1, \ldots, n$, such that

$$\sum_{i=1}^{n} P(B_i) = 1$$

and

$$p(B_i B_j) = 0$$

for all i and j (figure 6.1(*b*)). Then

$$p(A) = \sum_{i=1}^{n} p(AB_i) = \sum_{i=1}^{n} p(A|B_i)p(B_i)$$

and we have

$$p(B_j|A) = \frac{p(B_j)p(A|B_j)}{\sum_{i=1}^{n} p(B_i)p(A|B_i)}.$$

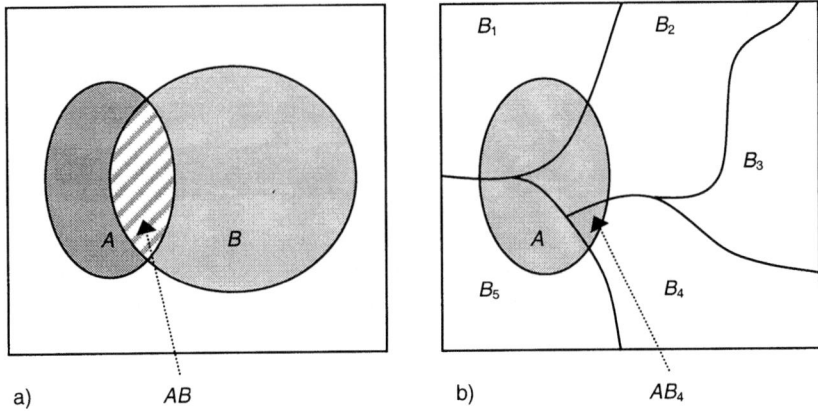

Figure 6.1. Schematic representation of the probability of propositions.

Example. Consider two machines in a production line. The productivity of machine 1 is twice that of machine 2, but machine 1 produces three times as many non-conforming products as machine 2. What is the probability that a product picked at random and found to be non-conforming was produced by machine 1?

To solve this problem, let B_i be the proposition 'the product was produced by machine i' and A the proposition 'the product is non-conforming'. Then $p(B_1) = 2p(B_2)$ and $p(A|B_1) = 3p(A|B_2)$. Therefore, from BT,

$$p(B_1|A) = \frac{2 \times 3}{2 \times 3 + 1} = \frac{6}{7}.$$

Simple as it is, BT is the basis of the theory of *Bayesian inference* [3,4,26]. The richness of this subject appears when we apply the theorem in a time context, taking B in (6.1) as a proposition whose *prior* probability $p(B)$ is modified into the *posterior* probability $p(B|A)$ *after* the occurrence of A has been observed. In this sense, BT can be considered as a mathematical description of the process of learning from experience.

6.2 BT in terms of probability density functions

In metrological applications, the propositions of interest usually take the forms:

- A: data d about a given quantity Q are observed in the course of the measurement experiment;
- B: the value of Q lies within a given infinitesimal interval $(q, q + dq)$.

Therefore, we write
$$p(B) = f(q)\,dq$$
and
$$p(B|A) = f(q|d)\,dq$$
where $f(q)$ and $f(q|d)$ are the pdfs that describe the states of knowledge about Q prior and posterior to measurement, respectively. Of course, in general these pdfs are not equal. Strictly, their difference should be emphasized by writing the posterior, for example, as $g(q|d)$ or $f_{Q|d}(q|d)$. This distinction is normally not necessary, since it is implied by the arguments of the respective functions.

The probabilities $p(A)$ and $p(A|B)$ are a little more difficult to interpret. Let us, for the moment, write these probabilities formally as
$$P(A) = f(d)\,dd$$
and
$$P(A|B) = f(d|q)\,dd.$$
Thus, from BT we have
$$f(q|d)\,dq = \frac{f(q)\,dq\, f(d|q)\,dd}{f(d)\,dd}$$
or, cancelling the differentials,
$$f(q|d) = Kf(d|q)f(q)$$
where
$$K = \frac{1}{f(d)}. \tag{6.2}$$
Also, from the normalization condition (3.2), we have
$$\int_{-\infty}^{\infty} f(q|d)\,dq = K \int_{-\infty}^{\infty} f(d|q)f(q)\,dq = 1 \tag{6.3}$$
and therefore, comparing (6.2) and (6.3)
$$f(d) = \int_{-\infty}^{\infty} f(d|q)f(q)\,dq.$$

Consider now the meaning of $f(d|q)$. There would be no problem in interpreting this function as a pdf in the variable d given that $Q = q$ if d referred to the possible values of a quantity D. But this is not so: d stands for *given general data* about Q, or even about other quantities related to Q. Therefore, we have to think of $f(d|q)$ as the probability density of the data, now considered as variables, *if* the value of Q was known to be equal to q, in a sense, 'forgetting' that d is given. For this, a *probability model* is required to interpret the data.

Afterwards, once we have the expression for $f(d|q)$ in terms of both q and d, we return to the original meaning of these symbols, the former as a variable and the latter as a given value or values. Note that the probability model is a concept quite distinct from that of the *measurement model* that has been discussed hitherto.

It is then more convenient to rewrite the argument of the function $f(d|q)$ as $q|d$, to emphasize that q is the variable and d is a given value or values. Since that function is *not* a pdf in q, we will write $f(d|q) = l(q|d)$ and will call it the *likelihood* of the data d. BT then takes the form

$$f(q|d) = Kl(q|d)f(q) \qquad (6.4)$$

where

$$K = \left[\int_{-\infty}^{\infty} l(q|d)f(q)\,dq \right]^{-1}.$$

It is clear from this expression that the likelihood can be multiplied by a constant or, more generally, by any function of d alone, since that constant or function would cancel out in (6.4). For this reason, the likelihood does not need to be normalized and should not be interpreted as a pdf in neither of its arguments. Moreover, since the constant K can be found from the normalization condition of the posterior pdf $f(q|d)$, BT will normally be written in the form 'posterior is proportional to likelihood times prior' or

$$f(q|d) \propto l(q|d)f(q).$$

In this way, there is no need to worry at all about the meaning of the (unconditional) density $f(d)$.

6.3 Example

These ideas should become clearer through the following example. Suppose that the state of knowledge about a certain quantity Q is represented by a normal pdf centred at $\dot{q}_e$ and standard deviation $\dot{u}_q$, that is,

$$f(q) = \frac{1}{\sqrt{2\pi}\,\dot{u}_q} \exp\left[-\frac{1}{2} \frac{(q - \dot{q}_e)^2}{\dot{u}_q^2} \right].$$

Suppose now that an unbiased method of experimental measurement is available and that an observation of Q is made by this method, yielding the value q_e. From experience it is known that any single observation made by the method can be assumed to be drawn from a normal frequency distribution of values, centred at the unknown value q of Q and known variance u_q. Therefore, the likelihood is

$$l(q|q_e, u_q) \propto \exp\left[-\frac{1}{2} \frac{(q - q_e)^2}{u_q^2} \right] \qquad (6.5)$$

Example   171

and from BT, the posterior state of knowledge becomes

$$f(q|q_e, u_q) \propto \exp\left[-\frac{1}{2}\frac{(q-q_e)^2}{u_q^2} - \frac{1}{2}\frac{(q-\dot{q}_e)^2}{\dot{u}_q^2}\right]. \tag{6.6}$$

The right-hand side of (6.6) can be written in a more convenient form using the following reasoning. We first collect the terms in the exponent that multiply equal powers of q:

$$f(q|q_e, u_q) \propto \exp\left[-\frac{1}{2}\left(\frac{1}{u_q^2} + \frac{1}{\dot{u}_q^2}\right)q^2 + \left(\frac{q_e}{u_q^2} + \frac{\dot{q}_e}{\dot{u}_q^2}\right)q - \frac{1}{2}\left(\frac{q_e^2}{u_q^2} + \frac{\dot{q}_e^2}{\dot{u}_q^2}\right)\right].$$

Next we note that the term

$$\exp\left[-\frac{1}{2}\left(\frac{q_e^2}{u_q^2} + \frac{\dot{q}_e^2}{\dot{u}_q^2}\right)\right]$$

does not depend on the variable q and can thus be lumped into the proportionality constant of $f(q|q_e, u_q)$. Therefore, without any loss of generality we write

$$f(q|q_e, u_q) \propto \exp\left[-\frac{1}{2}\left(\frac{1}{u_q^2} + \frac{1}{\dot{u}_q^2}\right)q^2 + \left(\frac{q_e}{u_q^2} + \frac{\dot{q}_e}{\dot{u}_q^2}\right)q\right]. \tag{6.7}$$

We now define

$$\frac{1}{\ddot{u}_q^2} = \frac{1}{u_q^2} + \frac{1}{\dot{u}_q^2}$$

and

$$\frac{\ddot{q}_e}{\ddot{u}_q^2} = \frac{q_e}{u_q^2} + \frac{\dot{q}_e}{\dot{u}_q^2}$$

and multiply the left-hand side of (6.7) by the constant

$$\exp\left[-\frac{1}{2}\frac{\ddot{q}_e^2}{\ddot{u}_q^2}\right].$$

The posterior density becomes

$$f(q|q_e, u_q) \propto \exp\left[-\frac{1}{2}\frac{(q-\ddot{q}_e)^2}{\ddot{u}_q^2}\right] \tag{6.8}$$

i.e. a normal pdf centred at the posterior best estimate $\ddot{q}_e$ and standard deviation $\ddot{u}_q$. Note that, without any need for further calculation, it may be concluded that the proportionality constant in (6.8) is equal to $(\sqrt{2\pi}\ddot{u}_q)^{-1}$. This analysis shows the convenience of postponing the calculation the proportionality constant until the final expression of the posterior pdf has been obtained.

172   Bayesian inference

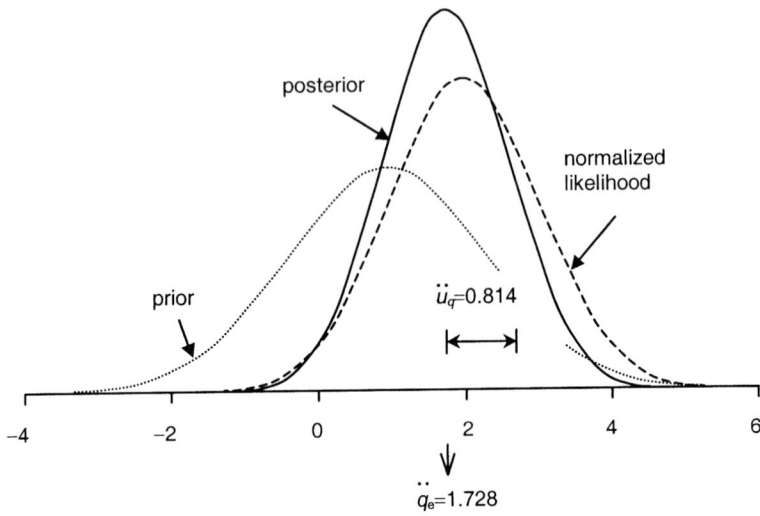

Figure 6.2. The posterior pdf with normal prior pdf and normal likelihood.

Example. Figure 6.2 illustrates the appearance of the posterior given a prior with $\dot{q}_e = 1$ and $\dot{u}_q^2 = 1.96$. The likelihood is a normal with expectation $q_e = 2$ and a smaller variance $u_q^2 = 1$. Because of this, the likelihood 'dominates', in the sense that the posterior becomes closer to the likelihood than to the prior. Specifically, in this example we have $\ddot{q}_e = 1.728$ and $\ddot{u}_q = 0.814$. If the prior and likelihood were to be interchanged, the same posterior would result with a now dominant prior. Note that the posterior variance is smaller than those of the prior and of the likelihood, indicating a more complete posterior state of knowledge. Even though all normalization constants are irrelevant, the likelihood has been drawn in its normalized form for clarity.

Consider now the following variant of the example: the likelihood is again given by (6.5) but the prior is a rectangular pdf over an interval of width $2b$ centred at $\dot{q}_e$. BT then immediately gives

$$f(q|q_e, u_q) = \frac{1}{K} \exp\left[-\frac{1}{2}\frac{(q-q_e)^2}{u_q^2}\right] \tag{6.9}$$

for $\dot{q}_e - b < q < \dot{q}_e + b$, where

$$K = \int_{\dot{q}_e-b}^{\dot{q}_e+b} \exp\left[-\frac{1}{2}\frac{(q-q_e)^2}{u_q^2}\right] dq.$$

In this case, the posterior variance $\ddot{u}_q$ needs to be obtained by numerical integration of the posterior density. The posterior pdf is symmetric only if

Non-informative priors   173

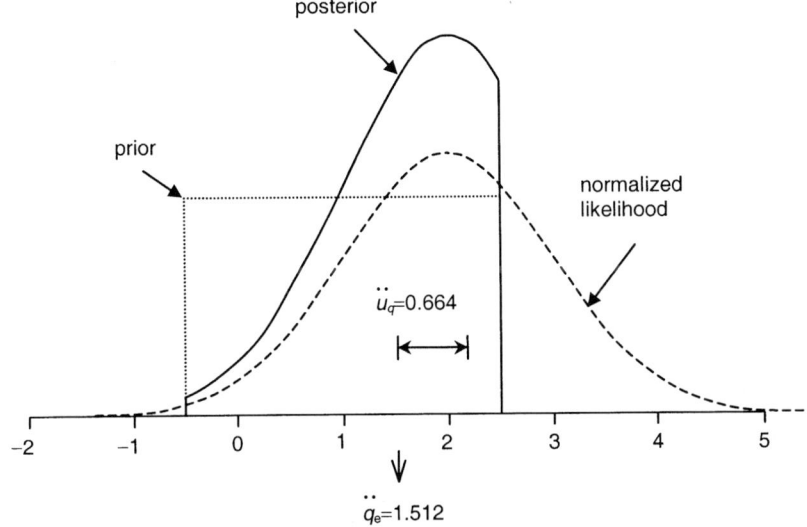

Figure 6.3. The posterior pdf with rectangular prior and normal likelihood.

$\dot{q}_e = q_e$, resulting in equal prior and posterior best estimates. Otherwise, $\ddot{q}_e$ has to be obtained numerically also.

Example. Figure 6.3 illustrates the shape of (6.9) for $\dot{q}_e = 1$, $b = 1.5$, $q_e = 2$ and $u_q = 1$, for which values $\ddot{q}_e = 1.512$ and $\ddot{u}_q = 0.664$ result.

6.4 Non-informative priors

In the previous examples, the posterior best estimate $\ddot{q}_e$ lies in between the prior best estimate $\dot{q}_e$ and the datum q_e. The exact location of $\ddot{q}_e$ depends on the size of the measurement uncertainty u_q, in the first case relative to the prior uncertainty $\dot{u}_q$ and, in the second case, relative to the half-width b. If u_q is small in comparison with $\dot{u}_q$ or b, the likelihood's normal shape 'dominates'.

Now, it is usually appropriate to conduct statistical inference as if no knowledge about the quantities of interest existed *a priori*. In this way, one would be sure to let the data 'speak for themselves' and not be influenced by a possibly dominant prior pdf that might appear to be subjective or not very well justified. This is called the 'jury principle'. We quote from [4]:

> cases are tried in a law court before a jury that is carefully screened so that it has no possible connections with the principals and the events of the case. The intention is clearly to ensure that information gleaned from 'data' or testimony may be assumed to dominate prior ideas that

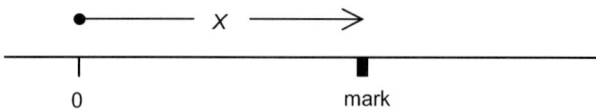

Figure 6.4. The position of a mark on a rail.

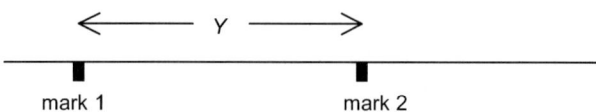

Figure 6.5. The separation of two marks on a rail.

members of the jury may have concerning the guilt or innocence of the defendant.

Consider, then, how a state of *total prior ignorance* can be represented [3, 4, 13, 25]. Suppose that we are interested in the position of a mark on a very long rail. Call X the distance of the mark to some arbitrary point, which we label as zero (figure 6.4). If the mark lies to the right of this point, we say that the quantity is positive (and negative otherwise). Assume that all we know is that the mark is there, but we are absolutely ignorant on where it lies. Represent this state of knowledge (or rather, ignorance) by the pdf $f(x)$, i.e. $f(x)\,dx$ is the probability that the mark lies in the interval $(x, x+dx)$. Now somebody comes along and tells us that the mark has been displaced some distance l, that is, if the previous location was at x, the new location is at $x' = x + l$, where l can be positive or negative. Let $f(x')\,dx'$ be the probability that the mark lies in the interval $(x', x'+dx')$. But since we are still ignorant about where the mark lies, we should assign equal probabilities to the mark being on either interval. Therefore we have $f(x)\,dx = f(x')\,dx'$ or, since $dx = dx'$, $f(x) = f(x') = f(x+l)$ for all values l. This is only possible if $f(x)$ is a constant, the so-called Bayes' prior.

Assume now that there are two marks on the rail. It does not matter *where* the marks are located, only their *separation* is of interest to us. Let this quantity be represented by Y and assume that nothing about it is known to us (figure 6.5). Define $f(y)\,dy$ as the probability that the separation is within the interval $(y, y+dy)$, $y > 0$. Now somebody takes a photograph of the marks and presents it to us, informing that the picture magnifies reality c times, i.e. if the actual separation is y, the separation in the picture is $y' = cy$, where c is an unknown positive number. Following the same reasoning as before, we have $f(y)\,dy = f(y')\,dy'$ or $f(y) = cf(cy)$ for all positive values c. This is only possible if $f(y)$ is proportional to $1/y$, the so-called Jeffreys' prior.

It is concluded that the representation of a state of total ignorance about a certain quantity Q should be described by one of two possible forms of a *noninformative prior*. The first form is a constant and applies to a quantity for which, if a set of observations about it were to be given, the addition of a constant l to all of them would change the best estimate from q_e to $q_e + l$. The second form of non-informative prior is proportional to $1/q$ and applies to a quantity for which, if a set of observations about it were to be given, multiplying all of them by a positive constant c would change the best estimate from q_e to cq_e.

Non-informative priors are discussed extensively in [3,4]. For our purposes, it suffices to assert that the question of whether Bayes' or Jeffreys' prior is more appropriate has to be decided on the basis of the likelihood, because it is this function that describes how the data are to be taken into account. As was previously mentioned, in most cases the likelihood is constructed according to the probability model that is assumed to hold for the given measurement problem. The probability model is basically a frequency distribution (called the *parent distribution* in [4]) that contains the quantity or quantities to be inferred as *variable parameters*. If some quantity appears as the expectation of the parent distribution, it is considered to be a *location parameter* and satisfies the criterion corresponding to Bayes' prior. Similarly, if a quantity appears as the variance or the standard deviation of the parent distribution, with *positive* possible values, it is considered to be a *scale parameter* and satisfies the criterion corresponding to Jeffreys' prior. In other words, whether a quantity should be regarded as a location or a scale parameter does not depend on the quantity itself, but on the context in which it appears in the probability model.

Now, obviously, neither non-informative priors are proper pdfs, because they cannot be normalized. However, this is of no *practical* importance, because the dominance of the likelihood will, in most cases, yield a normalizable posterior pdf.

6.5 Repeated measurements

In this section we consider the following problem: a quantity Q is measured independently n times, the results are the values $\mathbf{q} = (q_1 \quad \ldots \quad q_n)^{\mathrm{T}}$. No prior knowledge about Q is available. It is desired to obtain the corresponding best estimate and standard uncertainty.

We have already addressed the conventional solution to this problem: it was given in section 3.8 and was called TA analysis. We shall now present two Bayesian alternatives, taken from [32]. The first applies to the case in which the values $\mathbf{q}$ are assumed to be sampled from a normal probability model. It will be studied in section 6.5.1. In the second alternative, presented in section 6.5.2, a rectangular probability model is assumed. The treatment in section 6.5.3 concerns a series of normally distributed values that have been converted to digital indications. Finally, in section 6.5.4, we consider the case where, in addition to the values $\mathbf{q}$, the corresponding variances or error bounds are known.

176   Bayesian inference

6.5.1 Normal probability model

Whenever a set of observations **q** about Q is given, one should try to dwell on the physical reasons that lead to the appearance of different values. If the experimental evidence leads us to believe that the deviations of the q_is from the (true) value of the quantity are caused by several random effects and that these deviations are independent of one another, it is reasonable to think of the data as being drawn independently from an infinite population of values whose *frequency* distribution is normal, centred at the unknown value of the quantity Q and having an unknown variance V.

From the Bayesian point of view, Q and V are considered as being two measurands, only the first of which is identified with a *physical* quantity. The relationship between Q and V does not arise from the measurement model, but from the probability model. The quantity V is normally not of interest in itself, but its existence conditions what we can infer about Q. In Bayesian theory, a quantity such as V is called a *nuisance parameter*.

Conditional on given values q of Q and v of V, the pdf for any given datum q_i follows from the probability model and is

$$f(q_i|q,v) \propto \frac{1}{v^{1/2}} \exp\left[-\frac{1}{2}\frac{(q_i-q)^2}{v}\right].$$

In the context of BT, this density is interpreted as being proportional to the likelihood function $l(q,v|q_i)$. Since the observations are assumed to be independent, the global likelihood $l(q,v|\mathbf{q})$ is obtained by multiplying the individual likelihoods. Moreover, since Q and V are also independent, and they act as location and scale parameters, respectively, their joint prior separates into the product of a constant (Bayes' prior) for the former multiplied by v^{-1} (Jeffreys' prior) for the latter. The joint posterior becomes

$$f(q,v|\mathbf{q}) \propto \frac{1}{v^{n/2+1}} \exp\left[-\frac{1}{2v}\sum_{i=1}^{n}(q_i-q)^2\right].$$

It is convenient to rearrange this expression using the following identity

$$\sum_{i=1}^{n}(q_i-q)^2 = (n-1)s^2 + n(q-\bar{q})^2$$

where

$$\bar{q} = \frac{1}{n}\sum_{i=1}^{n} q_i \qquad (6.10)$$

and

$$s^2 = \frac{1}{n-1}\sum_{i=1}^{n}(q_i-\bar{q})^2. \qquad (6.11)$$

Then

$$f(q, v|\mathbf{q}) \propto \frac{1}{v^{n/2+1}} \exp\left[-\frac{1}{2}\frac{(n-1)s^2 + n(q-\bar{q})^2}{v}\right]. \quad (6.12)$$

Note that in this analysis $\bar{q}$ and s^2 are just *shorthand notation*; these parameters should *not* be interpreted here as estimates of the mean and variance of the parent distribution.

Inference about Q

Applying the marginalization rule of probability theory, the information pertaining to Q may be now decoded from the joint posterior by 'integrating out' the nuisance parameter V. For this we use the known definite integral

$$\int_0^\infty x^\alpha \exp(-ax^\beta)\,dx \propto \frac{1}{\beta a^{(\alpha+1)/\beta}} \quad (6.13)$$

where $\alpha > -1$, $\beta > 0$ and $a > 0$. After some rearrangement and due replacement of symbols we obtain

$$f(q|\mathbf{q}) \propto [(n-1)s^2 + n(q-\bar{q})^2]^{-n/2}. \quad (6.14)$$

The expectation and variance of this pdf can be obtained by numerical integration. However, it is much more convenient to perform the change of variable

$$T = \frac{\sqrt{n}}{s}(Q - \bar{q})$$

because then, by letting t represent the possible values of T, we obtain

$$f(t) \propto \left(1 + \frac{t^2}{n-1}\right)^{-n/2}$$

which is Student's-t density with $n-1$ degrees of freedom, see section 4.1.3. The expectation of this density is equal to zero and, from (4.4), its variance is

$$r^2 = \frac{n-1}{n-3}.$$

From properties (3.5) and (3.6) we obtain the expectation and variance of Q. The results are

$$q_e = \bar{q}$$

and

$$u_q^2 = \frac{(rs)^2}{n}.$$

Note that in this analysis n does not need to be 'large'; it is only required that the number of observations be at least two. However, n must be greater than three in order for the standard uncertainty to exist.

The expanded uncertainty is evaluated using the fact that a symmetric coverage interval for the variable T about $t = 0$ is bound by the t-quantiles $t_{p,n-1}$. Therefore, the symmetric coverage interval for Q is $(\bar{q} - t_{p,n-1}s/\sqrt{n}, \bar{q} + t_{p,n-1}s/\sqrt{n})$, i.e. the same as that obtained with conventional TA analysis. However, in Bayesian analysis the coverage factor is $k_p = t_{p,n-1}/r$.

Example. Recall the second example in section 3.8. A length Q is measured $n = 5$ times; the results are (215, 212, 211, 213, 213) nm. These values give $\bar{q} = 212.8$ nm, $s = 1.48$ nm and $r = 2$. The Bayesian standard uncertainty is 1.33 nm, while the TA standard uncertainty is 0.66 nm. The 95 % coverage interval turns out to be (210.96, 214.64) nm using either TA or Bayesian analysis.

Inference about V

Although the best estimate and associated uncertainty corresponding to the quantity V will rarely be needed, they can nevertheless be derived easily. In effect, it is only necessary to integrate the joint density (6.12) over q. Thus,

$$f(v|\mathbf{q}) \propto \int_{-\infty}^{\infty} \frac{1}{v^{n/2+1}} \exp\left\{-\frac{1}{2v}[(n-1)s^2 + n(q-\bar{q})^2]\right\} dq$$

$$\propto \frac{v^{1/2}}{v^{n/2+1}} \exp\left[-\frac{(n-1)s^2}{2v}\right] \int_{-\infty}^{\infty} \frac{1}{v^{1/2}} \exp\left[-\frac{n(q-\bar{q})^2}{2v}\right] dq.$$

Since the integrand in the second expression is a normal pdf, the integration yields a constant. Therefore

$$f(v|\mathbf{q}) \propto \frac{1}{v^{(n+1)/2}} \exp\left[-\frac{(n-1)s^2}{2v}\right]. \tag{6.15}$$

This density is less well known; it corresponds to that labelled as 'inverse chi-squared' in [26]. From the expressions for the expectation and variance given in that reference for this density, we conclude that the best estimate of the variance is

$$v_e = (rs)^2$$

with a standard uncertainty equal to the square root of

$$u_v^2 = \frac{2(rs)^4}{n-5}.$$

Note that u_v is the uncertainty associated with the best estimate of the variance, and should, by no means, be taken as the 'uncertainty of the uncertainty'. To obtain u_v, at least six values q_i are needed.

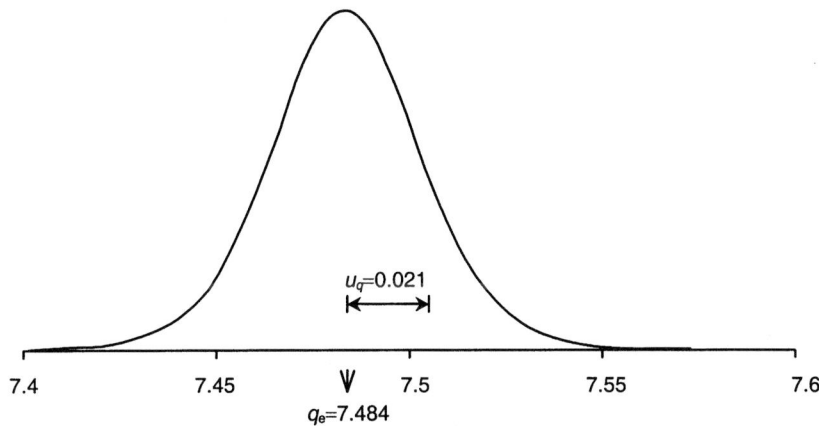

Figure 6.6. The pdf for Q with normal probability model. The parameters are $n = 10$ and $s^2 = 0.003\,492$.

Example. Consider again the example in section 3.8. A quantity Q is measured repeatedly 10 times. The resulting numeric values are:

$$7.489$$
$$7.503$$
$$7.433$$
$$7.549$$
$$7.526$$
$$7.396$$
$$7.543$$
$$7.509$$
$$7.504$$
$$7.383$$

and the unit is mm. These values give $q_e = \bar{q} = 7.484$ mm, $s^2 = 0.003\,492$ mm^2 and $r^2 = 1.2857$. The TA procedure gives $u_q = 0.019$ mm. The Bayesian method produces instead $u_q = 0.021$ mm. The latter also gives $v_e = 0.004\,489$ mm^2 and $u_v = 0.002\,839$ mm^2. Figures 6.6 and 6.7 depict the pdfs (6.14) and (6.15), respectively, for the values in this example.

180　　　Bayesian inference

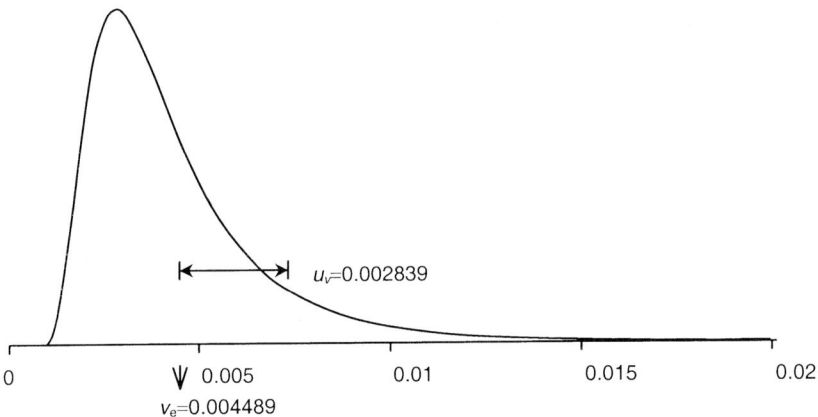

Figure 6.7. The pdf for V with normal probability model. The parameters are $n = 10$ and $s^2 = 0.003\,492$.

6.5.2 Rectangular probability model

Suppose now that the normal probability model used in the previous subsection is not realistic. For example, assume that the value of Q cannot be taken as fixed, because it is known that it varies randomly between two limits that are not perfectly known. If the time interval between measurements is larger than the characteristic fluctuation time, the values $\mathbf{q}$ may be considered to be sampled from a rectangular parent distribution centred at Q and having a half-width W, the latter being a nuisance (scale) parameter.

Conditional on given values q of Q and w of W, the pdf for each datum q_i, considered as a variable, is rectangular in the interval $(q-w, q+w)$. Accordingly, its likelihood is proportional to $1/w$ over this interval and is zero otherwise. The multiplication of all the individual likelihoods yields a global likelihood proportional to $1/w^n$ valid for $q - w < q_1, \ldots, q_n < q + w$. This last condition is equivalent to $q - w < q_a \leq q_b < q+w$, where $q_a = \min\{\mathbf{q}\}$ and $q_b = \max\{\mathbf{q}\}$.

With a constant prior for Q and a $1/w$ prior for W, the joint posterior becomes

$$f(q, w | \mathbf{q}) \propto \frac{1}{w^{n+1}}$$

for $q - w < q_a$ and $q_b < q + w$.

Inference about Q

To integrate out the nuisance parameter W, we need to establish its bounds. Given the values q, q_a and q_b, the conditions $q - w < q_a$ and $q_b < q + w$ imply that $q - q_a < w$ and $q_b - q < w$. Therefore, w does not have an upper bound, while its lower bound is equal to the maximum between $q - q_a$ and $q_b - q$. Denoting this lower bound as w_m we have

$$f(q|\mathbf{q}) \propto \int_{w_m}^{\infty} \frac{dw}{w^{n+1}} \propto \frac{1}{w_m^n}$$

or

$$f(q|\mathbf{q}) \propto \begin{cases} \dfrac{1}{(q-q_a)^n} & \text{if } q > \tilde{q} \\ \dfrac{1}{(q_b-q)^n} & \text{if } q < \tilde{q} \end{cases} \qquad (6.16)$$

where

$$\tilde{q} = \frac{q_a + q_b}{2}.$$

Since this density is symmetric about $\tilde{q}$ we conclude that

$$q_e = \tilde{q}.$$

The variance is a little more tedious to obtain. For this we need the proportionality constant K of the density, obtained from the normalization condition. The result is

$$K = \frac{n-1}{2}\left(\frac{\delta}{2}\right)^{n-1}$$

where $\delta = q_b - q_a$.

We now proceed to apply (3.3) to obtain the variance. After integration and some algebra we get

$$u_q^2 = \frac{\delta^2}{2(n-2)(n-3)}$$

for which calculation the number of measurements must be greater than three.

Inference about W

To infer an estimate for W we recognize that, given the values w, q_a and q_b, the conditions $q - w < q_a$ and $q_b < q + w$ imply that $q_b - w < q < w + q_a$. Therefore, integrating the joint posterior $f(q,w|\mathbf{q})$ over the variable q in this interval yields

$$f(w|\mathbf{q}) \propto \frac{2w - \delta}{w^{n+1}} \qquad (6.17)$$

for $w > \delta/2$. This density is not symmetric. Both its expectation and its variance are obtained by first deriving the proportionality constant. The calculations are easy, though they require a bit of algebra. The results are

$$w_e = \frac{n}{n-2}\left(\frac{\delta}{2}\right)$$

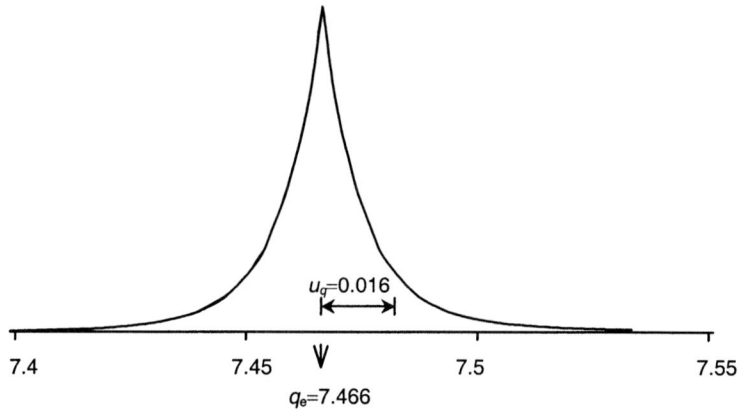

Figure 6.8. The cusp pdf for Q with rectangular probability model. The parameters are $q_a = 7.383$, $q_b = 7.549$ and $n = 10$.

and

$$u_w^2 = \frac{2n}{(n-2)^2(n-3)} \left(\frac{\delta}{2}\right)^2$$

valid for $n > 3$.

Example. Consider once more the list of values in the previous example. Under the assumptions in this section, we have $q_a = 7.383$ mm, $q_b = 7.549$ mm and $\delta = 0.166$ mm, from which $q_e = \tilde{q} = 7.466$ mm, $u_q = 0.016$ mm, $w_e = 0.104$ mm and $u_w = 0.018$ mm. Figures 6.8 and 6.9 depict the pdfs (6.16) and (6.17), respectively, that correspond to the values in this example. Because of its shape, the former may be termed the 'cusp' pdf.

6.5.3 Digital indications

Suppose now that the indications are produced by a digital display resolution device having a digital step equal to δ. The values **q** are then of the form $q_i = k_i \delta$ where the k_is are integers, see section 3.9 and figure 3.24. If the analogue, pre-converted values are assumed to proceed from a normal frequency distribution $N(x)$ centred at the unknown value q and variance v (as in section 6.5.1), the probability of observing the value $q_i = k_i \delta$ is [12]

$$l(q, v|q_i) = \int_{(k_i - 1/2)\delta}^{(k_i + 1/2)\delta} N(x)\,dx$$
$$= \Phi(b_i) - \Phi(a_i)$$

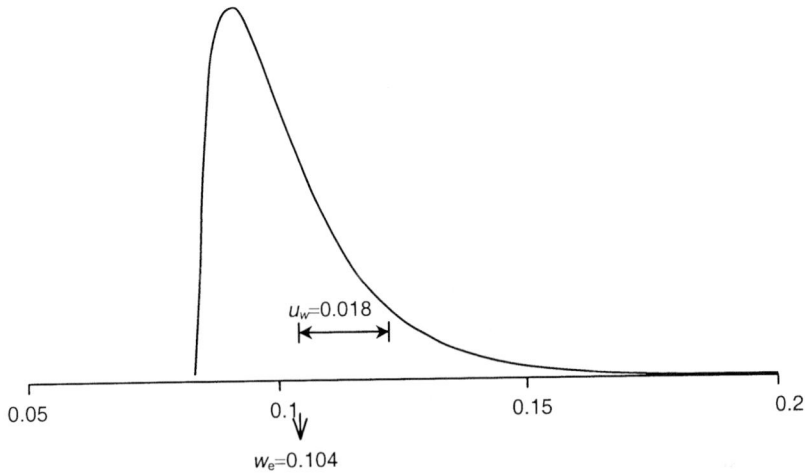

Figure 6.9. The pdf for W with parameters $n = 10$ and $\delta = 0.166$.

where

$$a_i = \frac{(k_i - 1/2)\delta - q}{\sqrt{v}}$$

$$b_i = \frac{(k_i + 1/2)\delta - q}{\sqrt{v}}$$

and Φ is the standard normal cumulative function, see figure 6.10. Thus, with the joint prior $f(q, v) \propto 1/v$, the joint posterior becomes

$$f(q, v|\mathbf{q}) \propto \frac{1}{v} \prod_{i=1}^{n}[\Phi(b_i) - \Phi(a_i)].$$

The best estimate q_e and the standard uncertainty u_q are then found as the expectation and standard deviation of the pdf

$$f(q|\mathbf{q}) \propto \int_0^\infty \prod_{i=1}^{n}[\Phi(b_i) - \Phi(a_i)]\frac{dv}{v} \tag{6.18}$$

respectively. All integrals must be carried out numerically, although unfortunately this is not an easy task.

Example. Table 6.1, taken from [12], shows the number of times n_i that the indications $q_i = i\delta$ were observed in the course of the measurement of a quantity Q. The procedure in this subsection yields $q_e = 3.57\delta$ and $u_q = 0.13\delta$. The conventional procedure in section 3.9 also yields $q_e = 3.57\delta$ but a larger uncertainty, $u_q = 0.31\delta$. See figure 6.11 for the pdf (6.18) that corresponds to this example.

184   Bayesian inference

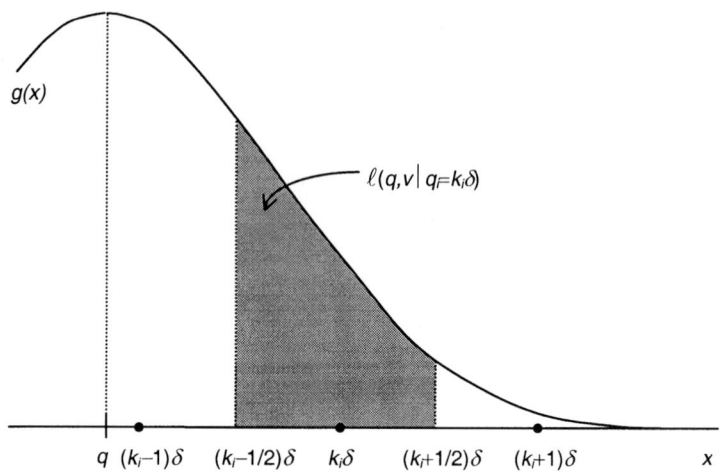

Figure 6.10. The likelihood of a digital indication $q_i = k_i\delta$, where k_i is an integer and δ is the digital step.

Table 6.1. The number of times n_i that the digital indications $q_i = i\delta$ were observed in the course of a measurement.

i	1	2	3	4	5	6
n_i	2	6	18	29	7	1

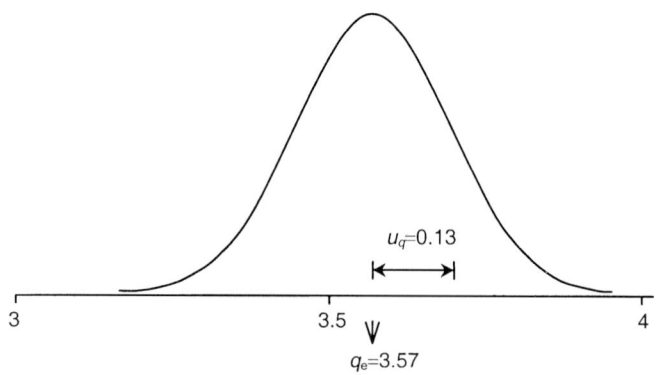

Figure 6.11. The pdf for a series of digital indications.

6.5.4 Reproducible measurements

In this subsection we address the problem of intercomparison measurements already considered in section 5.8.1. It will be shown to be an extension of the

Repeated measurements 185

example in section 6.3. Suppose that a quantity Q in the form of a material measure is circulated among n laboratories. Each laboratory reports a best estimate q_i and a standard uncertainty u_i. Which pdf best describes the pool of information $\mathbf{q} = (q_1 \ \ldots \ q_n)^T, \mathbf{u} = (u_1 \ \ldots \ u_n)^T$?

It is reasonable to assume that the values q_i reported by each laboratory are drawn from normal frequency distributions centred at the unknown value q of Q and variances u_i^2. Therefore, the likelihood associated with the first laboratory is

$$l(q|q_1, u_1) \propto \exp\left[-\frac{1}{2}\frac{(q-q_1)^2}{u_1^2}\right]. \tag{6.19}$$

From this expression it is evident that Q acts as a location parameter, to which there corresponds a non-informative constant prior. Thus, $l(q|q_1, u_1)$ is also proportional to $f(q|q_1, u_1)$. The latter becomes the prior density for the second laboratory, whose data likelihood can also be taken as a normal. (Evidently, the way the laboratories are ordered is irrelevant.) Therefore, as in (6.6), we have

$$f(q|q_1, q_2, u_1, u_2) \propto \exp\left[-\frac{1}{2}\frac{(q-q_1)^2}{u_1^2} - \frac{1}{2}\frac{(q-q_2)^2}{u_2^2}\right].$$

Thereafter we proceed sequentially to arrive at the final posterior pdf

$$f(q|\mathbf{q}, \mathbf{u}) \propto \exp\left[-\frac{1}{2}\sum_{i=1}^{n}\frac{(q-q_i)^2}{u_i^2}\right].$$

Following the same reasoning that led to (6.8), it is easy to verify that this posterior is a normal centred at

$$q_e = \left(\sum_{i=1}^{n}\frac{1}{u_i^2}\right)^{-1}\sum_{i=1}^{n}\frac{q_i}{u_i^2} \tag{6.20}$$

and variance

$$u_q^2 = \left(\sum_{i=1}^{n}\frac{1}{u_i^2}\right)^{-1}. \tag{6.21}$$

The reader should note that, with a different notation, these expressions are exactly equal to (5.42) and (5.43), respectively.

Example. Re-interpret the list of numeric values in the two previous examples as intercomparison measurements with standard uncertainties as shown in table 6.2. From (6.20) and (6.21) we get $q_e = 7.506$ and $u_q = 0.014$ (figure 6.12).

186   Bayesian inference

Table 6.2. Intercomparison measurement results.

i	q_i	u_i
1	7.489	0.059
2	7.503	0.045
3	7.433	0.085
4	7.549	0.023
5	7.526	0.057
6	7.396	0.069
7	7.543	0.071
8	7.509	0.031
9	7.504	0.082
10	7.383	0.046

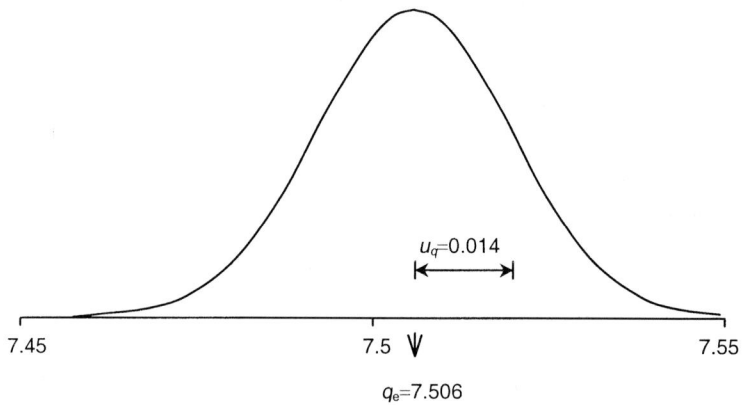

Figure 6.12. The pdf corresponding to a series of intercomparison measurements.

Assume now that, instead of reporting an uncertainty u_i, each laboratory gives an error bound b_i about the best estimate q_i, claiming a rectangular pdf over the interval $(q_i - b_i, q_i + b_i)$. We use a non-informative constant prior for $f(q)$ and a rectangular probability model for each likelihood, i.e. $f(q|q_1, b_1)$ is constant in the interval $(q_1 - b_1, q_1 + b_1)$. Proceeding sequentially in this way, we find

$$f(q|\mathbf{q}, \mathbf{b}) = \frac{1}{q_b - q_a} \qquad (6.22)$$

where

$$q_a = \max\{q_i - b_i\}$$
$$q_b = \min\{q_i + b_i\}$$

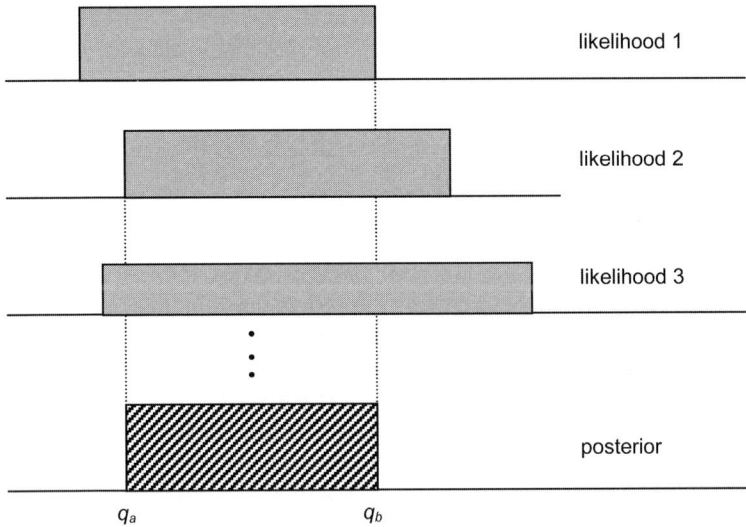

Figure 6.13. Overlapping intervals reported by the laboratories participating in an intercomparison.

and $\mathbf{b} = (b_1 \ \ldots \ b_n)^T$. Since the pdf (6.22) is rectangular, the posterior results are

$$q_e = \frac{q_a + q_b}{2}$$

and

$$u_q^2 = \frac{(q_b - q_a)^2}{12}.$$

Note that the posterior pdf makes sense only if $q_b > q_a$, which means that all intervals $(q_i - b_i, q_i + b_i)$ must overlap (figure 6.13).

Example. Re-interpret the standard uncertainties in table 6.2 as error bounds. Then $q_a = 7.526 > q_b = 7.429$ and the list would be inconsistent.

6.6 The principle of maximum entropy (PME)

It was argued in section 6.4 that it is usually appropriate to use BT ignoring any knowledge that might exist *a priori* about the quantities of interest. However, if prior information was *always* disregarded, one of the greatest advantages of the theorem would be lost.

For example, we might be given a prior best estimate q_e about some quantity Q, together with the standard uncertainty u_q. Now, suppose that whoever actually calculated the values q_e and u_q did not tell us the procedure he or she used. Only the best estimate and the standard uncertainty are conveyed to us, nothing else. Which pdf would best represent *our* state of knowledge? In other words, which pdf possesses q_e as its expectation and u_q^2 as its variance? The answer is, of course, not unique: virtually an unlimited number of pdfs can be envisaged to agree with that information.

This is a very serious problem: if we intend to use one of these pdfs as a prior in BT the consequent posterior will be influenced by our choice. For example, if we pick a dominant prior, any additional data about Q would be almost irrelevant. This issue was addressed in [25], where it was pointed out that, to be objective, the choice of prior should be 'maximally non-committal' with regard to the given information, that is, it should depend solely on that information, not on some arbitrary and personal point of view. This desideratum is, in fact, the 'jury principle', and it was used in the previous derivation of the two forms of non-informative priors.

Despite its relevance, it is hardly appropriate here to delve too deeply into the answers that have been given to solve this problem. A rather convincing approach is that adopted by the subject of *information theory*, initiated by Shannon and Weaver in 1949 [47], where all pdfs that comply with a given set of restrictions are regarded as a class of functions to each of which an *information measure* is attached. Because of its strong similarity with the corresponding term in the field of statistical mechanics, this measure was called the *entropy* of the pdf. It is defined as

$$S = -\int f(q) \ln[f(q)]\, dq. \tag{6.23}$$

The function to be selected is that which maximizes the entropy, thus the name *principle of maximum entropy* (PME) by which this method is known. Two applications will be illustrated here. Before doing so, however, it must be pointed out that the number S as an information measure applies only to the pdfs of a quantity that behaves as location parameter. To see why this is so, assume that we transform the quantity Q into Q', that is, we write $q' = g(q)$. This implies that the pdf $f(q)$ will be transformed into $f(q')$. The new entropy becomes

$$S' = -\int f(q') \ln[f(q')]\, dq'.$$

Now, if the transformation is not to alter our state of knowledge about either Q or Q', we must have $f(q)\, dq = f(q')\, dq'$. Therefore,

$$S' = -\int f(q) \ln\left[f(q)\frac{dq}{dq'}\right] dq.$$

and we conclude that, in order for $f(q)$ and $f(q')$ to have the same information measure, i.e. $S = S'$, the transformation must be such that $dq' = dq$ or $q' = q + l$, where l is any constant value. As we saw earlier, such non-informative transformation applies to a quantity that can be regarded as a location parameter.

6.6.1 Range and estimate given

The following problem is briefly mentioned in Clause 4.3.8 of the *Guide*: consider a quantity Q whose value is contained within the known range (q_a, q_b). The best estimate q_e is also known, but this value is not necessarily at the centre of the interval. We seek a function $f(q)$ for which S as defined by (6.23) is a maximum and is such that it complies with the following constraints:

$$\int_{q_a}^{q_b} f(q)\,dq = 1$$

and

$$\int_{q_a}^{q_b} q f(q)\,dq = q_e.$$

The solution turns out to be

$$f(q) = A e^{\lambda q}$$

where A and λ are to be found from the constraints. It should be stressed that this pdf, as well as any others that are obtained by the PME, *should not* be interpreted a statements of the statistical behaviour of the indicated values of Q.

To obtain the parameters A and λ it is convenient to perform the change of variable

$$X = \frac{2Q - q_a - q_b}{q_b - q_a} \tag{6.24}$$

such that $-1 \leq x \leq 1$. Then the pdf of X is of the same form as that of Q and we have for its corresponding parameters A and λ the system

$$\int_{-1}^{1} f(x)\,dx = \frac{A}{\lambda}(e^\lambda - e^{-\lambda}) = 1$$

and

$$\int_{-1}^{1} x f(x)\,dx = \frac{A}{\lambda}\left[e^\lambda\left(1 - \frac{1}{\lambda}\right) + e^{-\lambda}\left(1 + \frac{1}{\lambda}\right)\right] = x_e$$

where

$$x_e = \frac{2q_e - q_a - q_b}{q_b - q_a}.$$

Once A and λ are found (numerically), we apply equation (3.10)

$$u_x^2 = \int_{-1}^{1} x^2 f(x)\,dx - x_e^2$$

with

$$\int_{-1}^{1} x^2 f(x)\,dx = \frac{A}{\lambda}\left[e^\lambda\left(1 - \frac{2}{\lambda} + \frac{2}{\lambda^2}\right) - e^{-\lambda}\left(1 + \frac{2}{\lambda} + \frac{2}{\lambda^2}\right)\right].$$

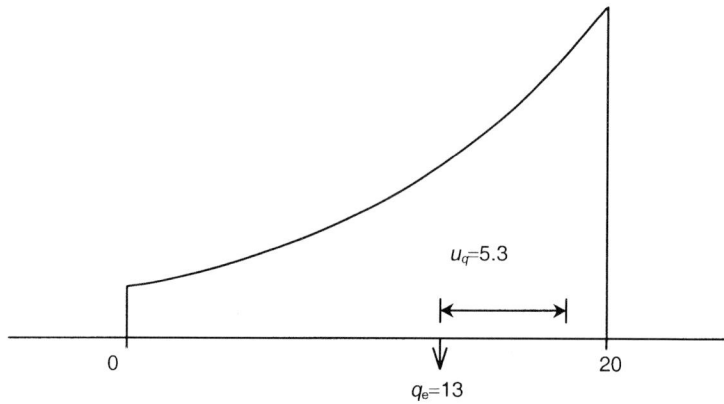

Figure 6.14. The maximum entropy pdf $f(q) = A \exp(\lambda q)$ corresponding to values of Q in the interval $(0, 20)$ and known best estimate $q_e = 13$.

The uncertainty associated with the best estimate of Q is finally obtained as

$$u_q = \frac{q_b - q_a}{2} u_x.$$

It should be noted that if the best estimate q_e is at the centre of the interval (q_a, q_b) then $x_e = 0$ and it follows by inspection that the normalization and expectation constraints are satisfied for $\lambda = 0$ and $A = 0.5$. Therefore, the maximum entropy pdf corresponding to this situation is rectangular. This gives strong theoretical support to an otherwise intuitive result.

Example. The value of an adimensional quantity Q is known to be contained within the interval $(0, 20)$; its best estimate is $q_e = 13$, therefore $x_e = 0.3$. Results are $\lambda = 0.9531$, $A = 0.4316$, $u_x = 0.5296$ and $u_q = 5.3$. Figure 6.14 depicts the shape of the maximum entropy pdf that corresponds to this example.

6.6.2 Range, estimate and uncertainty given

Suppose now that, in addition to knowing that the value of Q lies in the range (q_a, q_b) and that the best estimate is equal to a given value q_e, the standard uncertainty u_q is also given. The maximum entropy pdf is in this case

$$f(q) = A \exp(\lambda_1 q + \lambda_2 q^2)$$

where the constants A, λ_1 and λ_2 are to be found from the constraints

$$\int_{q_a}^{q_b} f(q) \, dq = 1$$

$$\int_{q_a}^{q_b} q f(q) \, dq = q_e \tag{6.25}$$

and

$$\int_{q_a}^{q_b} q^2 f(q) \, dq = u_q^2 + q_e^2. \tag{6.26}$$

This system of equations is rather difficult to solve numerically, because the constraint (6.26) involves the integral of $q^3 \exp(\lambda_2 q^2)$ that does not have a closed-form expression.

A simplified situation, considered in [55], is that where q_e is at the centre of the interval. The pdf must then be symmetric, which is only possible if $\lambda_1 = 0$. We then write $\lambda_2 = \lambda$ and

$$f(x) = A \exp(\lambda x^2)$$

for X defined by the transformation (6.24). The constraint (6.25) in terms of $f(x)$ is automatically satisfied for $x_e = 0$ and we have

$$\int_{-1}^{1} f(x) \, dx = 2A \int_{0}^{1} \exp(\lambda x^2) \, dx = 1$$

and

$$\int_{-1}^{1} x^2 f(x) \, dx = 2A \int_{0}^{1} x^2 \exp(\lambda x^2) \, dx = u_x^2.$$

The elimination of the constant A gives

$$g(\lambda) = u_x^2 \tag{6.27}$$

where

$$g(\lambda) = \left(\int_{0}^{1} \exp(\lambda x^2) \, dx \right)^{-1} \int_{0}^{1} x^2 \exp(\lambda x^2) \, dx.$$

It is easy to see that the range of the function $g(\lambda)$ is the interval $(0, 1)$. This implies that a solution to (6.27) is possible only if $u_x < 1$.

We now proceed inversely, that is, we choose different values of λ and substitute them in (6.27) to obtain a value for u_x^2. Thus, choosing $\lambda = 0$ gives $u_x^2 = \frac{1}{3}$. This corresponds to a rectangular pdf. Positive values of λ give $\frac{1}{3} < u_x^2 < 1$ and correspond to a U-shaped pdf, while negative values of λ give $0 < u_x^2 < \frac{1}{3}$ and correspond to a truncated normal pdf (figure 6.15).

As noted in [55], the problem just considered can be formulated from an opposite point of view. Suppose, as before, that we know Q to be contained within the interval (q_a, q_b) with best estimate q_e at the midpoint. However, instead of knowing the standard uncertainty u_q, we have a reasonable estimate for the ratio r of the mid-point probability $f(q_e) \, dq$ to the probability at the limits of the interval

192   Bayesian inference

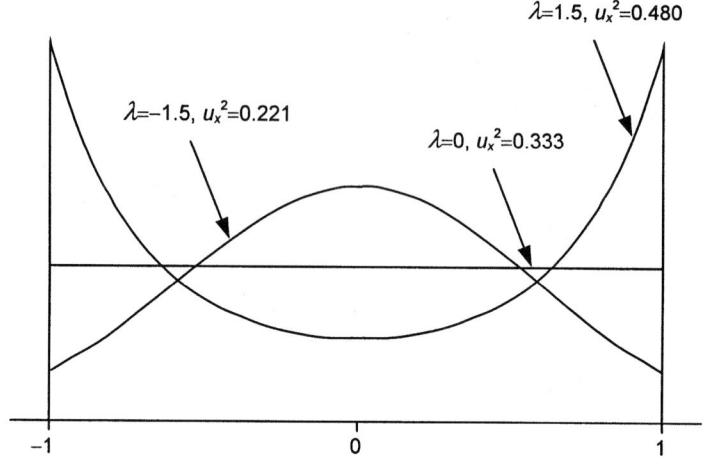

Figure 6.15. The maximum entropy pdf $f(x) = A \exp(\lambda x^2)$ corresponding to values of X in the interval $(-1, 1)$, known best estimate $x_e = 0$ and known variance u_x^2.

$f(q_a) \, dq = f(q_b) \, dq$. In terms of the variable X, this is equivalent to having $f(x) = A \exp(\lambda x^2)$ and knowing the ratio

$$r = \frac{f(0)}{f(1)} = \frac{1}{e^\lambda}.$$

The value $\lambda = -\ln r$ is then replaced in (6.27) to obtain an uncertainty that is consistent, in the maximum entropy sense, with the available information. For example, for $r = 2$ we get $u_x = 0.525$.

6.6.3 Estimate and uncertainty given

If the interval (q_a, q_b) is not given as part of the information, it must be assumed that the range of possible values of Q extends from $-\infty$ to ∞. The given estimate q_e is then automatically 'centred'. Whatever the value u_q, the normal is the maximum entropy pdf. This conclusion is in complete agreement with the pdf that one would otherwise tend to use intuitively. For example, it validates the assumption for the likelihoods (6.5) and (6.19).

6.7 Evaluation of the producer's and consumer's risks

For a measurand Q, the right and left coverage uncertainties were defined in section 4.2 as the distances from the best estimate q_e to the limits of the coverage interval I_p, where p is the coverage probability (equal to the integral of $f(q)$ over I_p). If the pdf is symmetric and the standard uncertainty exists, the coverage factor k_p was defined as the ratio of the coverage uncertainty to the standard uncertainty; in this case the *Guide*'s expanded uncertainty U_{pq} becomes identical to our definition of coverage uncertainty.

The expanded uncertainty is mostly useful for decision purposes, for example, when products are tested for compliance with a certain specification. Let Q be the quantity of interest of a given item in the production line, μ the desired or nominal value and ε the specification limit. The specification interval is $S = (\mu - \varepsilon, \mu + \varepsilon)$. However, according to the criterion in [23], products are accepted only if the estimated value q_e falls within the acceptance interval $A_p = (\mu - \kappa_p \varepsilon, \mu + \kappa_p \varepsilon)$, where κ_p is the guardband factor, see figure 4.5 on page 111. This factor is calculated as

$$\kappa_p = 1 - \frac{U_{pq}}{\varepsilon} \qquad (6.28)$$

where it is assumed that $\varepsilon > U_{pq}$ (otherwise the acceptance interval is not defined).

As discussed in section 4.3, the consumer's and producer's risks appear in this context. The former is defined as the probability of false acceptance, that is, as the probability that the best estimate q_e is within A_p while the true (unknown) value of Q is outside S. The producer's risk is the probability of false rejection, that is, the probability of q_e being outside A_p while the true value of Q is within S. By following [33] we now proceed to evaluate these risks, first from the Bayesian viewpoint and next using the sampling theory approach.

6.7.1 The Bayesian viewpoint

In the following, the guardband factor will be written without a subscript to simplify notation. Let the state of knowledge about Q given the estimate q_e be represented by the pdf $f(q|q_e)$. With κ given, the consumer's and producer's risks *for the particular measured item* become, respectively,

$$R_C(q_e) = \int_{-\infty}^{\mu-\varepsilon} f(q|q_e) \, dq + \int_{\mu+\varepsilon}^{\infty} f(q|q_e) \, dq$$

if $|q_e - \mu| < \kappa \varepsilon$ and

$$R_P(q_e) = \int_{\mu-\varepsilon}^{\mu+\varepsilon} f(q|q_e) \, dq$$

if $|q_e - \mu| > \kappa \varepsilon$ (see figures 6.16 and 6.17).

To evaluate these risks, assume that in addition to q_e, the standard uncertainty u_q is also a datum. If nothing else is known about Q, the PME advises us to take

$$f(q|q_e) = \frac{1}{\sqrt{2\pi} u_q} \exp\left[-\frac{1}{2} \frac{(q - q_e)^2}{u_q^2}\right] \qquad (6.29)$$

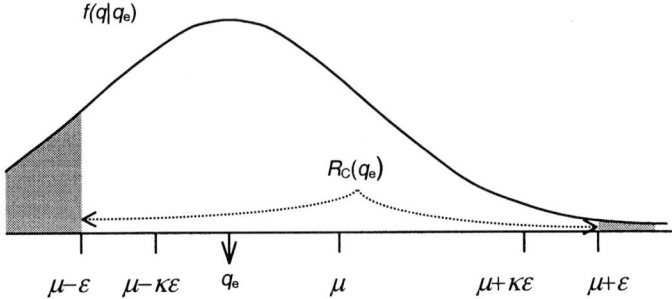

Figure 6.16. The consumer's risk for a measured item in the production line.

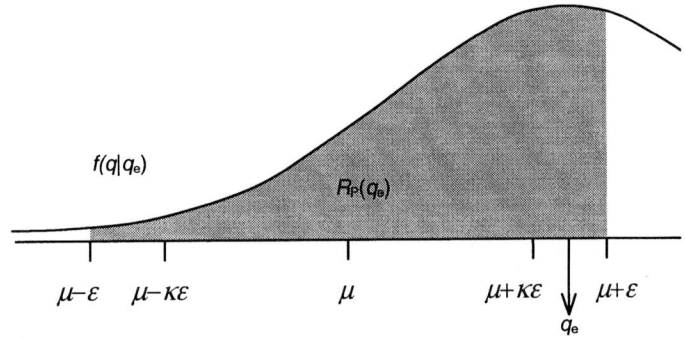

Figure 6.17. The producer's risk for a measured item in the production line.

such that

$$R_C(q_e) = 1 + \Phi(\delta - \nu) - \Phi(\delta + \nu) \quad \text{for } |\delta| < \kappa \nu \quad (6.30)$$

and

$$R_P(q_e) = \Phi(\delta + \nu) - \Phi(\delta - \nu) \quad \text{for } |\delta| > \kappa \nu \quad (6.31)$$

where

$$\nu = \frac{\varepsilon}{u_q}$$

$$\delta = \frac{\mu - q_e}{u_q}$$

and $\Phi(z)$ is the standard normal cumulative function, see section 4.1.1 and table A1 in the appendix.

The uncertainty u_q characterizes the measurement process; it can be obtained as the experimental standard deviation of several measurements of a *single* representative product, under repeatability conditions. The latter standard deviation, however, should *not* be divided by the number of measurements, as u_q is the standard uncertainty associated with a single measurement result.

The consumer's and producer's risks can also be defined *for the production line* as the expected proportions of falsely accepted and rejected items. These proportions are easily evaluated under the assumption that u_q is independent of the measured values q_e and that the *frequency* distribution $f(q_e)$ of these values is known experimentally. We then have

$$R_C = \int_{\mu-\kappa\varepsilon}^{\mu+\kappa\varepsilon} R_C(q_e) f(q_e) \, dq_e$$

and

$$R_P = \int_{-\infty}^{\mu-\kappa\varepsilon} R_P(q_e) f(q_e) \, dq_e + \int_{\mu+\kappa\varepsilon}^{\infty} R_P(q_e) f(q_e) \, dq_e.$$

To evaluate these expressions we take

$$f(q_e) = \frac{1}{\sqrt{2\pi}\sigma_q} \exp\left[-\frac{1}{2}\frac{(q_e - \mu)^2}{\sigma_q^2}\right] \tag{6.32}$$

where σ_q is the standard deviation of the measured values of the population of products; it can be estimated as the experimental standard deviation of the results obtained from the measurement of *several* products. Note that (6.32) has nothing to do with the PME; the chosen form of the frequency distribution $f(q_e)$ stems from the usual assumption of normality for a characteristic that varies randomly because of several non-systematic influences that cannot be rigorously controlled. We then have

$$R_C = \frac{1}{2\pi} \int_{-\kappa\rho}^{\kappa\rho} \left[\int_{-\infty}^{-\tau(\rho+y)} e^{-x^2/2} \, dx + \int_{\tau(\rho-y)}^{\infty} e^{-x^2/2} \, dx \right] e^{-y^2/2} \, dy$$

and

$$R_P = \frac{1}{2\pi} \int_{-\infty}^{-\kappa\rho} \int_{-\tau(\rho+y)}^{\tau(\rho-y)} e^{-(x^2+y^2)/2} \, dx \, dy$$
$$+ \frac{1}{2\pi} \int_{\kappa\rho}^{\infty} \int_{-\tau(\rho+y)}^{\tau(\rho-y)} e^{-(x^2+y^2)/2} \, dx \, dy$$

where

$$\tau = \frac{\sigma_q}{u_q}$$

and

$$\rho = \frac{\varepsilon}{\sigma_q}.$$

Results of the numerical integration of these equations are given in tables A3 and A4 in the appendix.

Example. It is desired to inspect a line of workpieces whose specification limit, set from functional considerations, is $\varepsilon = 10$ (arbitrary units). Assume that the value $u_q = 1$ applies, such that $\nu = 10$ and with a coverage factor $k = 2$ the value $\kappa = 0.8$ is obtained for the guardband factor.

Suppose that the measurement of three particular items gives the values 7, 8 and 9, respectively, for the difference $q_e - \mu$. From (6.30) and (6.31) one obtains 0.1 % for the consumer's risk of the first item (table A3) and 84.1 % for the producer's risk of the third (table A4). The second item is found to be in the limit of the acceptance interval. For this item, the consumer's risk is 2.3 % if accepted and the producer's risk is 97.7 % if rejected.

Now let $\sigma_q = 5$, so that $\tau = 5$ and $\rho = 2$. Consider first the case of no guardbanding ($\kappa = 1$). From tables A3 and A4 we obtain 1.1 % and 0.7 % for the expected proportion of workpieces that would be incorrectly accepted and rejected, respectively, after testing a large number of products. If $\kappa = 0.8$ is selected, the consumer's risk decreases to a negligible value, but the producer's risk increases to 6.0 %.

6.7.2 The frequency-based viewpoint

A different approach to the problem considered in this section, based on the concepts of conventional statistics, was presented in [11] and has been used recently [43].

As before let 2ε be the width of the specification interval, centred at the nominal value μ, and let κ be the guardband factor. Redefine Q as the quantity of interest for the *population* of products and $f(q)$ as the frequency distribution of the actual (true) values. Suppose that a single measurement is performed on a number of products that share the same (unknown) value of Q, and let $f(q_e|q)$ be the frequency distribution of the measured values (alternatively, this distribution may be interpreted as that of repeated measurements performed on a single product). The proportion of those products that are accepted is then

$$p|q = \int_{\mu-\kappa\varepsilon}^{\mu+\kappa\varepsilon} f(q_e|q)\, dq_e.$$

However, if q is outside the specification interval, all these products should have been rejected (figure 6.18). The consumer's risk then becomes

$$R_C = \int_{-\infty}^{\mu-\varepsilon} (p|q) f(q)\, dq + \int_{\mu+\varepsilon}^{\infty} (p|q) f(q)\, dq.$$

A similar reasoning gives the producer's risk:

$$R_P = \int_{\mu-\varepsilon}^{\mu+\varepsilon} (1 - p|q) f(q)\, dq.$$

Evaluation of the producer's and consumer's risks

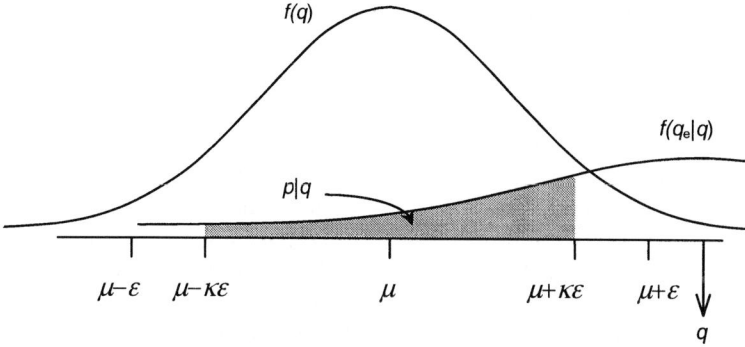

Figure 6.18. The frequency distribution $f(q_e|q)$ represents single measured values of items that share the same true value q. The shaded area represents the proportion of products that are accepted. But since q is outside the specification interval, all these products should have been rejected.

Parametrized expressions for these risks can be derived assuming that $f(q_e|q)$ and $f(q)$ are normal frequency distributions

$$f(q_e|q) = \frac{1}{\sqrt{2\pi}u_q} \exp\left[-\frac{1}{2}\frac{(q-q_e)^2}{u_q^2}\right] \qquad (6.33)$$

$$f(q) = \frac{1}{\sqrt{2\pi}\sigma} \exp\left[-\frac{1}{2}\frac{(q-\mu)^2}{\sigma^2}\right]$$

such that

$$R_C = \frac{1}{2\pi}\int_{-\infty}^{-\rho}\int_{-\tau(\kappa\rho+y)}^{\tau(\kappa\rho-y)} e^{-(x^2+y^2)/2}\,dx\,dy$$
$$+ \frac{1}{2\pi}\int_{\rho}^{\infty}\int_{-\tau(\kappa\rho+y)}^{\tau(\kappa\rho-y)} e^{-(x^2+y^2)/2}\,dx\,dy$$

and

$$R_P = \frac{1}{2\pi}\int_{-\rho}^{\rho}\left[\int_{-\infty}^{-\tau(\kappa\rho+y)} e^{-x^2/2}\,dx + \int_{\tau(\kappa\rho-y)}^{\infty} e^{-x^2/2}\,dx\right]e^{-y^2/2}\,dy$$

where now

$$\tau = \frac{\sigma}{u_q}$$

and

$$\rho = \frac{\varepsilon}{\sigma}.$$

(Note the difference between $f(q|q_e)$ in (6.29) and $f(q_e|q)$ in (6.33): their expressions are the same, yet the former is an assumed probability density on the variable q while the latter is a frequency distribution on the variable q_e.)

This approach has two important shortcomings. The first is that it does not allow one to evaluate the risks associated with accepting or rejecting a given measured item, i.e. no expressions analogous to (6.30) and (6.31) can be derived.

The second—and much more serious—shortcoming is that the variance σ^2 of the actual values of the population cannot be known. In [43] it is proposed that this variance be 'deconvolved' as

$$\sigma^2 = \sigma_q^2 - u_q^2 \qquad (6.34)$$

with the meaning of σ_q as in section 6.7.1. This equation, however, is not convincing, if only for the reason that u_q may in fact be *larger* than σ_q. This situation may occur, for example, if one uses an instrument whose uncertainty is to be attributed mainly to the comparatively bad resolution of its indicating device. It might happen that, if the products do not deviate much from their nominal value, all indications would be essentially the same, i.e. $\sigma_q = 0$.

In contrast, the Bayesian procedure does not need to assume anything about the frequency distribution of the true values.

Example. Consider one of the examples in [43], where $\sigma_q = \varepsilon/2$, $u_q = \varepsilon/8$ and $\kappa = 0.75$. The Bayesian model produces in this case $R_C = 0.06\%$ and $R_P = 8.2\%$. Instead, the frequency-based model leads to $R_C = 0.02\%$ and $R_P = 9.5\%$ after taking $\sigma^2 = 15\varepsilon^2/64$ from (6.34). This is not a very large *numerical* difference, but a very significant *conceptual* difference.

6.7.3 The risks in instrument verification

The consumer's and producer's risks also arise in connection with the work of calibration laboratories when an instrument is verified using a measurement standard. Let S denote the standard, G the gauged values and $Q = S - G$ the negative of the instrument's 'error'. With ε being a maximum acceptable error, the instrument is accepted if $q_e = s_e - g_e$ is found to be within the acceptance interval $(-\kappa\varepsilon, \kappa\varepsilon)$, but would have been rejected if the value of Q was known to be outside the specification interval $(-\varepsilon, \varepsilon)$. Given the value of S, the probability of the latter event follows from the pdf associated with G and is (figure 6.19)

$$p|s = \int_{-\infty}^{s-\varepsilon} f(g)\,dg + \int_{s+\varepsilon}^{\infty} f(g)\,dg$$

and since S may, in principle, take on any value, the consumer's risk becomes

$$R_C = \int_{-\infty}^{\infty} (p|s) f(s)\,ds \qquad \text{applicable if } |q_e| < \kappa\varepsilon. \qquad (6.35)$$

Evaluation of the producer's and consumer's risks

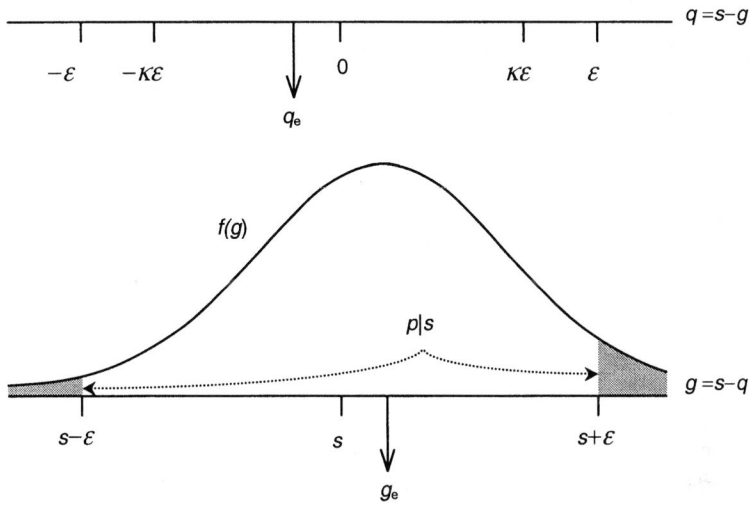

Figure 6.19. The value s is given and the indication g_e is obtained when the instrument is verified. The shaded areas represent the probability that the value of Q is outside the interval $(-\varepsilon, \varepsilon)$.

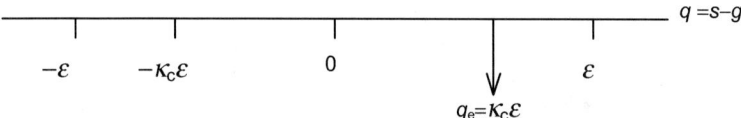

Figure 6.20. The critical guardband factor.

A similar line of reasoning gives the producer's risk:

$$R_P = \int_{-\infty}^{\infty} (1 - p|s) f(s) \, ds \qquad \text{applicable if } |q_e| > \kappa \varepsilon. \qquad (6.36)$$

Note that for a given value q_e, the critical guardband factor is $\kappa_c = |q_e|/\varepsilon$, such that if $\kappa < \kappa_c$ the instrument is rejected, and otherwise accepted (figure 6.20).

To evaluate the consumer's and producer's risks in instrument verification, assume that the pdfs associated with S and G are normal, i.e.

$$f(s) = \frac{1}{\sqrt{2\pi} u_s} \exp\left[-\frac{1}{2} \frac{(s-s_e)^2}{u_s^2}\right]$$

and

$$f(g) = \frac{1}{\sqrt{2\pi} u_g} \exp\left[-\frac{1}{2} \frac{(g-g_e)^2}{u_g^2}\right]$$

such that

$$R_C = 1 - \frac{1}{2\pi} \int_{-\infty}^{\infty} \int_{\tau(y-\rho)+\delta}^{\tau(y+\rho)+\delta} e^{-(x^2+y^2)/2} \, dx \, dy$$

for $|\delta|/(\tau\rho) < \kappa$ and

$$R_P = \frac{1}{2\pi} \int_{-\infty}^{\infty} \int_{\tau(y-\rho)+\delta}^{\tau(y+\rho)+\delta} e^{-(x^2+y^2)/2} \, dx \, dy$$

for $|\delta|/(\tau\rho) > \kappa$, where now

$$\delta = \frac{q_e}{u_g}$$

$$\tau = \frac{u_s}{u_g}$$

and

$$\rho = \frac{\varepsilon}{u_s}. \tag{6.37}$$

The results of the numerical integration of these equations are given in tables A5 and A6 in the appendix. Note that, strictly, the standard uncertainties u_s and u_g depend on the estimates s_e and g_e. However, this dependency was ignored in writing (6.35) and (6.36).

> **Example.** It is desired to verify the readings G of a digital multimeter (DMM). The manufacturer of the DMM claims that the absolute value of the 'error of measurement' $Q = |S-G|$ at its 10 V reading is less than $\varepsilon = 220$ μV. A multifunction electrical calibrator (MFC) that outputs a voltage S is supplied to the terminals of the DMM. With $s_e = 10$ V the indication $g_e = 9.999\,868$ V is obtained. Therefore, the error is $q_e = 132$ μV.
>
> Suppose that the standard uncertainties are $u_s = 11$ μV for the output of the MFC and $u_g = 55$ μV for the readings of the DMM. The integration parameters become $\delta = 2.4$, $\tau = \frac{1}{5}$ and $\rho = 20$. With no guardbanding ($\kappa = 1$) the DMM is accepted with a consumer's risk equal to 5.8 % (table A5). However, from (6.28) with $k = 2$ and $u_q = (u_s^2 + u_g^2)^{1/2} = 56$ μV, the value $\kappa = 0.5$ is obtained. Accordingly, the DMM is rejected with a producer's risk equal to 94.2 % (table A6).

It should be mentioned that, since the uncertainty of the measurement standard influences the probability of incorrectly accepting an unworthy instrument, some documentary standards recommend limiting its value. For example, in [22], one reads that 'the error attributable to calibration should be no more than one-third and preferably one-tenth of the permissible error of the confirmed equipment when in use'. Similarly, in [1] it is stated that 'the collective uncertainty of the measurement standards shall not exceed 25 % of the acceptable tolerance for each characteristic of the measuring and test equipment being calibrated or verified'. These statements should be interpreted as referring to the inverse of our parameter ρ, equation (6.37). The latter is sometimes called the 'TUR' for 'test uncertainty ratio'.

6.8 The model prior

Up to now, BT has been applied to just one quantity Q without much regard to its measurement model. In general, however, our interest lies in several quantities $\mathbf{Q}$, some or all of which are related explicitly through a set of model relations $\mathcal{M}(\mathbf{Q}) = \mathbf{0}$. Suppose that at the first stage of the evaluation process the state of knowledge about the quantities is assessed by the *joint* pdf $f(\mathbf{q})$, constructed in agreement with all information that is available up to that stage. After some additional data d about some or all the quantities $\mathbf{Q}$ is obtained, the posterior joint pdf is

$$f(\mathbf{q}|d) \propto l(\mathbf{q}|d) f(\mathbf{q}).$$

Because of the model relations, the quantities $\mathbf{Q}$ cannot be independent *a posteriori*. We can take into account this dependency by writing the prior pdf in the form

$$f(\mathbf{q}) \propto f_\mathrm{q}(\mathbf{q}) f_\mathcal{M}(\mathbf{q})$$

where $f_\mathrm{q}(\mathbf{q})$ might be called the *quantities prior* and $f_\mathcal{M}(\mathbf{q})$ the *model prior*. (The use of subscripts to differentiate between these two types of prior pdfs is here indispensable, since their arguments are the same.)

The quantities prior is constructed using whatever is known about the individual quantities. For example, if they are all independent, the quantities prior separates into the product of all the marginal pdfs, some or all of which may be non-informative.

In turn, the model prior is simply equal to zero if one of the relations is not fulfilled and is equal to one otherwise [52]. Thus, for models of the form $\mathcal{M}(\mathbf{Q}) = \mathbf{0}$, the model prior is Dirac's delta function

$$f_\mathcal{M}(\mathbf{q}) = \delta[\mathcal{M}(\mathbf{q})].$$

This function was defined in section 3.5.3 and was used there exactly in the sense of model prior.

Example. Consider the model $Q_1 = Q_2 Q_3$. If one knows that Q_1 is positive, that Q_2 is greater than a and that Q_3 is less than b, the model prior is

$$f_\mathcal{M}(\mathbf{q}) = \delta(q_1 - q_2 q_3) H(q_1) H(q_2 - a) H(b - q_3)$$

where $H(q)$ is the Heaviside unit step function, equal to one if $q \geq 0$ and to zero otherwise (figure 6.21).

202 Bayesian inference

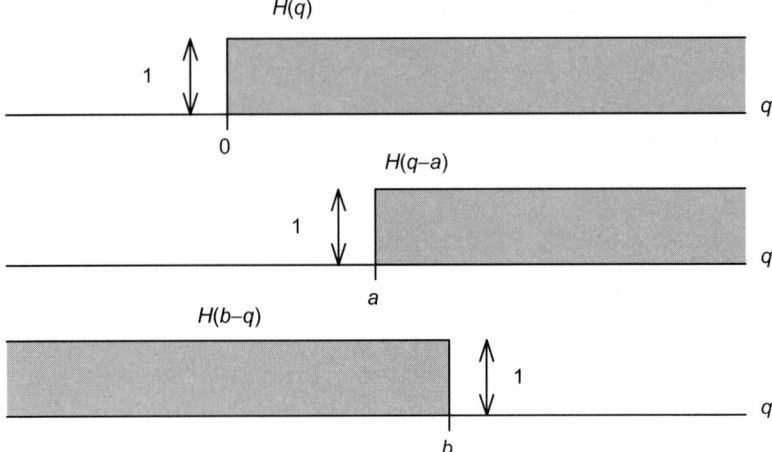

Figure 6.21. The Heaviside unit step function.

The model prior provides a very useful tool for determining posterior densities. We shall illustrate its application through some examples in the following section.

6.9 Bayesian analysis of the model $Z = \mathcal{M}(X, Y)$

As was discussed in section 3.4, the case of only one output quantity Z modelled explicitly as
$$Z = \mathcal{M}(X, Y)$$
is usually analysed by considering the linearized model
$$Z_L = a + c_x X + c_y Y \qquad (6.38)$$
where
$$a = \mathcal{M}(x_e, y_e) - (c_x x_e + c_y y_e)$$
and c_x and c_y are the sensitivity coefficients. One then assumes that in the neighbourhood of $\mathcal{M}(x_e, y_e)$ the quantities Z and Z_L behave very similarly and applies to the latter the LPU. In this way, u_z is obtained in terms of the standard and mutual uncertainties of the input quantities. This procedure applies strictly if the model is linear, and produces very reasonable values for u_z if the model is not too strongly nonlinear. However, as has already been mentioned, the LPU cannot be used if one desires to take into account any previous knowledge about Z that may be available.

Further, the methodology discussed in section 4.6 allows one to obtain the expanded uncertainty U_{pz} for a given coverage probability p. However, this procedure has many drawbacks, basically because it does not lead to a pdf $f(z)$ which can be used to derive the coverage interval.

Instead, the Bayesian analysis renders the pdf which most adequately expresses what is known about the output quantity including, if available, previous information about the same. From this pdf both the standard and expanded uncertainties may be derived. Thus, from BT we have

$$f(z, x, y|d) \propto l(z, x, y|d) f(z, x, y) \delta[z - \mathcal{M}(x, y)] \qquad (6.39)$$

where d represents the measurement data, $l(z, x, y|d)$ is the likelihood, $f(z, x, y)$ is the joint quantities prior and $\delta[z - \mathcal{M}(x, y)]$ is the model prior (this model prior assumes that there are no restrictions on the possible values of Z, X and Y, for example, that all or some of these quantities must be positive).

The pdf $f(z, x, y)$ depends on the type of prior knowledge about each of the quantities, the likelihood depends on the type of data that are obtained from measurement and the model prior determines the way by which the input quantities are to be integrated to obtain the posterior marginal pdf $f(z|d)$ for the output quantity.

6.9.1 The model $Z = X + Y$

The linear model $Z = X + Y$ is quite common. Among other situations, it applies in particular to a calibrand Z (either a material measure or the indications of a measuring instrument) that is calibrated using a measurement standard X and a comparator that measures the difference Y between Z and X. This model applies also to a quantity Z that is measured using an instrument whose indications are represented by the quantity Y and is subject to the influence of a constant systematic effect X.

Moreover, the additive model is also obtained from the more general model (6.38) by redefining Z, X and Y as $Z_L - a$, $c_x X$ and $c_y Y$, respectively. For these reasons, it deserves a close scrutiny.

In this subsection we assume that nothing is known about Z prior to the evaluation of the model. This situation is not very uncommon. In fact, as has already been mentioned many times, all the recommendations in the *Guide* are restricted to this situation. Furthermore, if X and Y are assumed to be independent, the data prior $f(z, x, y)$ is just equal to the product of the priors $f(x)$, $f(y)$ and $f(z)$, where the latter is a non-informative constant. Even though these priors may be fixed from measurement data, the symbol d in (6.39) refers to those data that are gathered *after* the prior $f(z, x, y)$ has been established. Therefore, if no measurements are made after $f(x)$ and $f(y)$ are available, the likelihood is equal to a constant and BT in its (6.4) form simply gives,

$$f(z, x, y) = f(x) f(y) \delta(x + y - z)$$

where $f(z, x, y)$ now refers to the posterior joint pdf, and the word 'posterior' refers not to the data (which does not exist) but to the state of knowledge after the model has been evaluated.

The posterior assessment for Z can now be obtained by integrating $f(z, x, y)$ over x and y (the order of integration is arbitrary). For this, the 'sifting' property of the delta function is used, see section 3.5.3. For example, integrating first over y one gets

$$f(z, x) = f_X(x) f_Y(z - x).$$

(The argument of the second pdf in the right-hand side of this equation no longer makes it clear that the pdf pertains to the quantity Y. Therefore, the use of subscripts is now mandatory.) Integrating next over x yields

$$f(z) = \int f_X(x) f_Y(z - x) \, dx. \tag{6.40}$$

We thus obtain a well-known result: the pdf of a quantity for which no prior knowledge exists and is modelled as the sum of two independent input quantities is equal to the *convolution* of the pdfs of the latter.

Convolutions are, in general, cumbersome to integrate. The procedure will be illustrated using rectangular prior pdfs for both X and Y, centred respectively at given values x_e and y_e, and having known (positive) half-widths e_x and e_y (figure 6.22). In other words, $f_X(x)$ will be taken as equal to $1/(2e_x)$ for x inside the interval $x_e \pm e_x$ and to zero outside this interval, while $f_Y(y)$ will be taken as equal to $1/(2e_y)$ for y inside the interval $y_e \pm e_y$ and to zero outside this interval. Therefore, in order for the integral in (6.40) not to vanish the following inequalities must be satisfied simultaneously:

$$x_{\min} < x < x_{\max}$$
$$y_{\min} < z - x < y_{\max}$$

where $x_{\max} = x_e + e_x$, $x_{\min} = x_e - e_x$ $y_{\max} = y_e + e_y$ and $y_{\min} = y_e - e_y$.

By adding these inequalities one gets the range over which $f(z)$ does not vanish: it is the interval $z_e \pm e_z$, where

$$z_e = x_e + y_e$$

and

$$e_z = e_x + e_y.$$

Next, the limits of the integral in (6.40) must be established. We call x_l the lower limit and x_u the upper limit. The previous inequalities require that

$$x_l = \begin{cases} z - y_{\max} & \text{if } z > z_e - e_x + e_y \\ x_{\min} & \text{otherwise} \end{cases}$$

and

$$x_u = \begin{cases} z - y_{\min} & \text{if } z > z_e + e_x - e_y \\ x_{\max} & \text{otherwise.} \end{cases}$$

Bayesian analysis of the model $Z = \mathcal{M}(X, Y)$

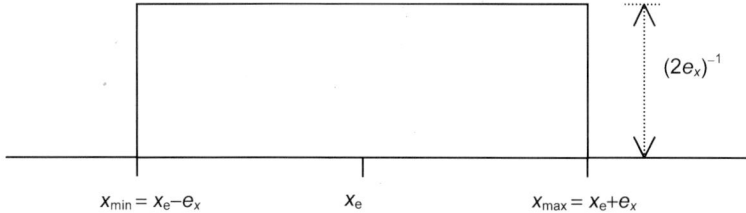

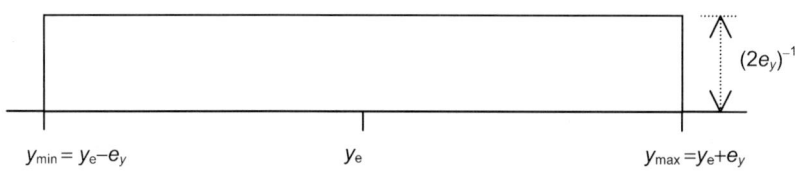

Figure 6.22. Rectangular pdfs for the input quantities.

These limits are applied to (6.40) with the integrand equal to the constant $c = 1/(4e_x e_y)$. The result depends on the relative values of e_x and e_y. For example, for $e_x < e_y$ one finds

$$f(z) = \begin{cases} c(z - z_e + e_z) & \text{for } z_e - e_z < z < z_e + e_x - e_y \\ 2ce_x & \text{for } z_e + e_x - e_y < z < z_e - e_x + e_y \\ c(z_e + e_z - z) & \text{for } z_e - e_x + e_y < z < z_e + e_z. \end{cases}$$

Figure 6.23 illustrates the shape of $f(z)$ for both $e_x < e_y$ and $e_y < e_x$. This pdf corresponds to the symmetrical trapezoidal pdf in section 3.3.2. Obviously, for $e_x = e_y$ the triangular pdf is obtained.

From equation (3.18), with

$$\phi = \frac{|e_x - e_y|}{e_z}$$

the square of the standard uncertainty becomes equal to

$$u_z^2 = \frac{e_x^2}{3} + \frac{e_y^2}{3} = u_x^2 + u_y^2.$$

This expression is hardly surprising; it could as well have been derived by applying the LPU to $Z = X + Y$. However, the fact that the Bayesian analysis renders the pdf for this quantity allows us to obtain the expanded uncertainty. The

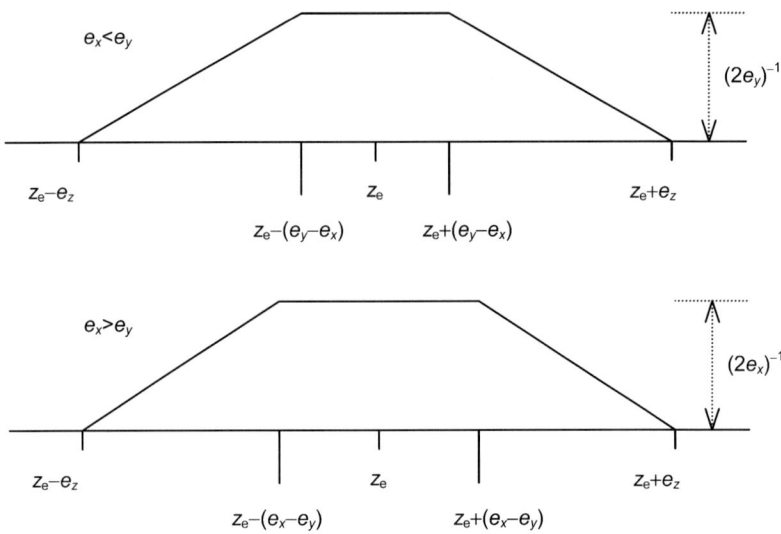

Figure 6.23. The trapezoidal pdf obtained as the result of a convolution of two rectangular pdfs.

following formula is easily obtained:

$$U_{pz} = \begin{cases} p\max(e_x, e_y) & \text{for } p < p_{\lim} \\ e_x + e_y - 2\sqrt{e_x e_y(1-p)} & \text{for } p > p_{\lim} \end{cases} \quad (6.41)$$

where

$$p_{\lim} = 1 - \min(e_x/e_y, e_y/e_x).$$

Let us compare this result with that obtained using the conventional procedure in section 4.6. From the WS formula, equation (4.7), and under the assumption that there is 'no uncertainty associated with the uncertainties u_x and u_y', the effective degrees of freedom are infinite. Therefore, in the conventional procedure one would use the normal pdf to obtain the expanded uncertainty.

Example. Consider $e_x = e_y = e$ and $p = 0.9545$. In the conventional procedure, this coverage probability is associated with a coverage factor $k = 2$. This leads to

$$\frac{U_{\text{Conv.}}}{e} = 2\sqrt{\frac{2}{3}} = 1.63.$$

Instead, the more appropriate Bayesian analysis, equation (6.41), gives

$$\frac{U_{\text{Bayes}}}{e} = 2\left(1 - \sqrt{1-p}\right) = 1.57.$$

6.9.2 The models $Z = X/Y$ and $Z = XY$

Consider now the model $Z = X/Y$. From its linearized version

$$Z_L = \frac{x_e}{y_e} + \frac{1}{y_e}X - \frac{x_e}{y_e^2}Y$$

we obtain

$$u_z^2 = \frac{1}{y_e^2}u_x^2 + \frac{x_e^2}{y_e^4}u_y^2$$

and conclude that, if $x_e = 0$, the uncertainty in knowing Y will have no effect on u_z.

Likewise, from the linearized version of the model $Z = XY$

$$Z_L = -x_e y_e + y_e X + x_e Y$$

we obtain

$$u_z^2 = y_e^2 u_x^2 + x_e^2 u_y^2$$

and conclude that, if x_e and y_e are both equal to zero, the uncertainty associated with the estimate $z_e = 0$ vanishes.

These obviously incorrect results are a consequence of linearizing the models at points where their curvature cannot be neglected. In this subsection it will be shown that Bayesian analysis leads quite naturally to the corresponding densities $f(z)$, from which the estimates and standard uncertainties are readily obtained.

Thus, to obtain the pdf of $Z = X/Y$ we write

$$f(z, x, y) = f_X(x) f_Y(y) \delta(z - x/y)$$

and let $w = x/y$. Integrating first over w and then over y yields

$$f(z = x/y) = \int f_X(zy) f_Y(y) |y| \, dy. \tag{6.42}$$

Similarly, to obtain the pdf of $Z = XY$ we use

$$f(z, x, y) = f_X(x) f_Y(y) \delta(z - xy)$$

and let $w = xy$. Integrating first over w and then over x yields

$$f(z = xy) = \int f_X(x) f_Y\left(\frac{z}{x}\right) \left|\frac{1}{x}\right| dx. \tag{6.43}$$

We shall illustrate, next, the integration of these equations by using rectangular priors for X and Y, as shown in figure 6.22. Thus, for the model

$Z = X/Y$ we set $x_e = 0$. Letting $y_{min} = y_e - e_y > 0$ and $y_{max} = y_e + e_y$, integration of equation (6.42) gives

$$f(z = xy) = \begin{cases} \dfrac{y_{max}^2 - y_{min}^2}{8e_x e_y} & \text{for } |z| < z_1 \\ \dfrac{1}{8e_x e_y}\left(\dfrac{e_x^2}{z^2} - y_{min}^2\right) & \text{for } z_1 < |z| < z_{max} \end{cases} \quad (6.44)$$

and

$$u_z^2 = \dfrac{1}{6}\dfrac{e_x^2}{e_y}\left(\dfrac{1}{y_{min}} - \dfrac{1}{y_{max}}\right)$$

where

$$z_1 = \dfrac{e_x}{y_{max}}$$

and

$$z_{max} = \dfrac{e_x}{y_{min}}.$$

Similarly, integrating (6.43) with rectangular priors for X and Y with $x_e = y_e = 0$ yields

$$f(z = xy) = \dfrac{1}{2e_x e_y}\ln\left(\dfrac{e_x e_y}{|z|}\right) \quad \text{for } |z| < e_x e_y \quad (6.45)$$

$$u_z = \dfrac{e_x e_y}{3}.$$

In both cases, $z_e = 0$ because of symmetry. Interestingly, the uncertainty of $Z = XY$ can be derived with recourse to the pdf $f(z = xy)$. Thus, for $Z = XY$ with X and Y independent and $E(X) = E(Y) = 0$ we have

$$u_z^2 = V(Z) = E(Z)^2 - E^2(Z) = E(X)^2 E(Y)^2 - 0 = u_x^2 u_y^2.$$

Figures 6.24 and 6.25 depict the pdfs (6.44) and (6.45), respectively. Note that the latter is a particular instance of the log pdf, equation (3.20).

6.9.3 The model $Z = X + Y$ with prior knowledge about Z

At the beginning of section 6.9.1 two typical applications of the model $Z = X+Y$ were discussed. In the first, Z is a quantity that is calibrated using a measurement standard X and a comparator that measures the difference Y between Z and X. In the second, Z is a quantity that is measured using an instrument whose indications are represented by the quantity Y and is subject to the influence of a constant systematic effect X. Either situation advises us to assume that [36]:

- knowledge about X is cast in the form of a pdf $f(x)$;
- direct measurement data d refer only to Y;

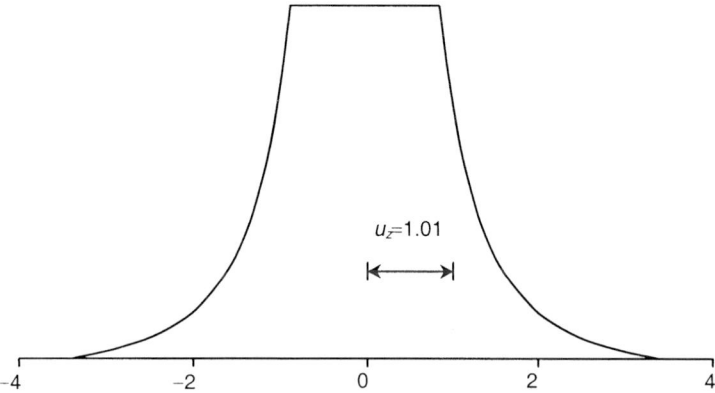

Figure 6.24. The pdf for $Z = X/Y$ assuming rectangular priors for X and Y with $x_e = 0$, $e_x = 7$, $y_e = 5$ and $e_y = 3$.

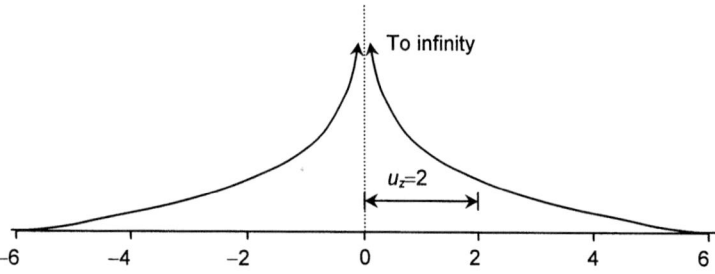

Figure 6.25. The pdf for $Z = XY$ assuming rectangular priors for X and Y with $x_e = y_e = 0$ and $e_x e_y = 6$.

- the data consist of the set of indicated values $y_1, \ldots, y_n$, $n \geq 2$;
- the data are drawn under repeatability conditions from a normal frequency distribution centred at the value of Y;
- nothing is known about Y before measurement;
- there is no correlation between X and Y; and
- previous knowledge about Z might exist independently.

Assume further that the prior information about X is available in either of two forms:

(i) the best estimate x_e and the standard uncertainty u_x are given; and
(ii) an interval of values (x_α, x_β) is given, inside which one is sure that the value of X lies.

According to the PME, the prior $f(x)$ corresponds to a normal density with expectation x_e and variance u_x^2 for case (i) and to a rectangular density centred at $x_e = (x_\alpha + x_\beta)/2$ with standard uncertainty $u_x = (x_\beta - x_\alpha)/\sqrt{12}$ for case (ii).

Similar considerations will be assumed to hold for the quantity Z. However, for this variable it is convenient to introduce the additional subscripts 'a' and 'b' to differentiate the situations after and before the measurement of Y, respectively. Thus, our goal is to obtain z_{ea} and u_{za} while z_{eb} and u_{zb} will be either given (normal prior) or calculated as $z_{eb} = (z_\alpha + z_\beta)/2$ and $u_{zb} = (z_\beta - z_\alpha)/\sqrt{12}$ (rectangular prior). Moreover, the case where nothing is known about Z before the measurement of Y will be considered as well. In this latter case we shall use Bayes' prior for $f(z)$, i.e. a non-informative constant.

With respect to the quantity Y, matters are much simplified if it is assumed that the variance of all indications y_i is equal to a known common value ϑ. In such case it was shown in [50] that if normal priors are assumed for both X and Z, the posterior pdf of Z is again a normal with expectation value and variance given by

$$z_{ea} = \left(\frac{x_e + \bar{y}}{u_x^2 + \sigma_y^2} + \frac{z_{eb}}{u_{zb}^2} \right) \left(\frac{1}{u_x^2 + \sigma_y^2} + \frac{1}{u_{zb}^2} \right)^{-1}$$

$$u_{za}^2 = \left(\frac{1}{u_x^2 + \sigma_y^2} + \frac{1}{u_{zb}^2} \right)^{-1}$$

respectively, where

$$\bar{y} = \frac{1}{n} \sum_{i=1}^{n} y_i$$

and

$$\sigma_y^2 = \vartheta/n$$

(this case is reviewed in section 6.9.4).

It is more realistic, however, to assume that the variance of the underlying frequency distribution of Y is not known; it will be treated as a nuisance parameter V. The prior $f(y)$ will be considered as a non-informative constant and the prior pdf for V will be taken as $f(v) \propto 1/v$. From (6.12), the posterior joint assessment for Y and V is

$$f(y, v|d) \propto h(y, v) \tag{6.46}$$

where

$$h(y, v) = \frac{1}{v^{1+n/2}} \exp\left[-\frac{1}{2} \frac{(n-1)s^2 + n(y - \bar{y})^2}{v} \right]$$

and

$$s^2 = \frac{1}{n-1} \sum (y_i - \bar{y})^2.$$

(Note again that, as in (6.10) and (6.11), $\bar{y}$ and s^2 are here shorthand notation; they have nothing to do with estimates of expectation and variance.)

Bayesian analysis of the model $Z = \mathcal{M}(X, Y)$ 211

Table 6.3. Prior knowledge about the quantities X and Z. 'Normal' refers to prior best estimate and standard uncertainty both known. 'Rectangular' means that the interval of possible values is known. 'Unknown' signifies complete absence of prior knowledge.

Case i	Quantities X	Z	Prior joint assessment $H_i(x, z)$
1	Normal	Unknown	$N(x; x_e, u_x)$
2	Rectangular	Unknown	$R(x; x_\alpha, x_\beta)$
3	Normal	Normal	$N(x; x_e, u_x) N(z; z_{eb}, u_{zb})$
4	Rectangular	Normal	$R(x; x_\alpha, x_\beta) N(z; z_{eb}, u_{zb})$
5	Normal	Rectangular	$N(x; x_e, u_x) R(z; z_\alpha, z_\beta)$
6	Rectangular	Rectangular	$R(x; x_\alpha, x_\beta) R(z; z_\alpha, z_\beta)$

The posterior joint assessment of the four quantities under consideration will be denoted as $f_i(x, y, z, v)$, where to save space the symbol '$|d$' (given d) has been suppressed and i denotes one of the cases in table 6.3 that summarizes the different combinations of prior knowledge. We then have

$$f_i(x, y, z, v) \propto h(y, v) H_i(x, z) \delta(x + y - z) \qquad (6.47)$$

where the joint assessment priors $H_i(x, z)$ are given in table 6.3. The notation in this table is

$$R(w; w_\alpha, w_\beta) = \begin{cases} 1 & \text{if } w_\alpha \leq w \leq w_\beta \\ 0 & \text{otherwise} \end{cases}$$

$$N(w; w_e, u_w) = \exp\left[-\frac{1}{2} \frac{(w - w_e)^2}{u_w^2}\right]$$

where w stands for x or z.

The information pertaining to a specific quantity may now be decoded from the joint posteriors (6.47) by 'integrating out' the other three quantities. We start by integrating the nuisance parameter V using (6.13). We then have, for $n \geq 2$,

$$f_i(x, y, z) \propto S(y) H_i(x, z) \delta(x + y - z) \qquad (6.48)$$

where

$$S(y) = \left[1 + \frac{1}{n-1} \frac{(y - \bar{y})^2}{\sigma^2}\right]^{-n/2}$$

and

$$\sigma^2 = \frac{s^2}{n}.$$

One may now obtain the posterior joint assessment for any two of the quantities X, Y and Z by integrating (6.48) over the third. The result shows the correlation between the two remaining quantities. In particular consider the posterior assessment of X and Y after integrating (6.48) over z. These quantities are independent *a priori*, but because of the model $Z = X + Y$ any prior information about the sum of X and Y will lead to their logical posterior correlation. For example, for case 3

$$f_3(x, y) \propto S(y) N(x; x_e, u_x) N(x + y; z_{eb}, u_{zb})$$

the right-hand side of which cannot be written as a product of a function of x and a function of y, as is the case for $f_1(x, y)$ and $f_2(x, y)$. From the expressions for $f_i(x, y)$ the posterior states of knowledge about X and Y may be obtained.

Since we are mainly interested in the posterior for Z we now integrate (6.48) over x and y (the order of integration is arbitrary). The results are

$$f_1(z) \propto \int_{-\infty}^{\infty} dx \, S(z - x) N(x; x_e, u_x)$$

$$f_2(z) \propto \int_{x_\alpha}^{x_\beta} dx \, S(z - x)$$

$$f_3(z) \propto f_1(z) N(z; z_{eb}, u_{zb})$$

$$f_4(z) \propto f_2(z) N(z; z_{eb}, u_{zb})$$

$$f_5(z) \propto f_1(z) R(z; z_\alpha, z_\beta)$$

$$f_6(z) \propto f_2(z) R(z; z_\alpha, z_\beta).$$

To facilitate the comparison of the results presented here, we select $x_\alpha = x_e - \sqrt{3} u_x$ and $x_\beta = x_e + \sqrt{3} u_x$ in order for the rectangular and normal prior densities for X to have the same expectation values and variances. Similarly, we take $z_\alpha = z_{eb} - \sqrt{3} u_{zb}$ and $z_\beta = z_{eb} + \sqrt{3} u_{zb}$. Moreover, it is convenient to cast the posteriors $f_i(z)$ into parametrized forms by means of the transformations

$$\begin{aligned} X' &= \frac{X - x_e}{u_x} \\ Z' &= \frac{Z - (x_e + \overline{y})}{(u_x^2 + \sigma^2)^{1/2}} \\ T' &= \frac{Z'(\eta^2 + 1)^{1/2} - X'}{\eta} \end{aligned} \quad (6.49)$$

where

$$\eta = \frac{\sigma}{u_x}.$$

Then, if we let x', z' and t' denote the possible values of X', Z' and T' and, in addition, define

$$F(t') = \left(1 + \frac{t'^2}{n-1}\right)^{-n/2}$$

Bayesian analysis of the model $Z = \mathcal{M}(X, Y)$

$$\mu' = \frac{z_{eb} - (x_e + \bar{y})}{(u_x^2 + \sigma^2)^{1/2}}$$

$$\sigma' = \frac{u_{zb}}{(u_x^2 + \sigma^2)^{1/2}}$$

$$z'_\alpha = \mu' - \sqrt{3}\sigma'$$

$$z'_\beta = \mu' + \sqrt{3}\sigma'$$

we have

$$f_1(z') \propto \int_{-\infty}^{\infty} dx' \, F(t') N(x'; 0, 1)$$

$$f_2(z') \propto \int_{-\sqrt{3}}^{\sqrt{3}} dx' \, F(t')$$

$$f_3(z') \propto f_1(z') N(z'; \mu', \sigma')$$

$$f_4(z') \propto f_2(z') N(z'; \mu', \sigma')$$

$$f_5(z') \propto f_1(z') R(z'; z'_\alpha, z'_\beta)$$

$$f_6(z') \propto f_2(z') R(z'; z'_\alpha, z'_\beta)$$

(note that t' is a function of both x' and z').

The posterior best estimates and standard uncertainties that correspond to these standardized probability densities can now be obtained. After evaluating the normalization constants, we have

$$z'_{ei} = \int_{-\infty}^{\infty} dz' \, z' f_i(z') \qquad (6.50)$$

$$u_{z'i}^2 = \int_{-\infty}^{\infty} dz' \, (z' - z'_{ei})^2 f_i(z') \qquad (6.51)$$

where the subscript 'a' has been dropped to simplify notation. From the relationship between Z and Z', the posterior best estimate of Z and associated standard uncertainty become

$$z_{ei} = z'_{ei}(u_x^2 + \sigma^2)^{1/2} + (x_e + \bar{y})$$

and

$$u_{zi} = u_{z'i}(u_x^2 + \sigma^2)^{1/2}.$$

Finally, the coverage probabilities are evaluated as

$$p_i = \int_{I_{pi}} dz' \, f_i(z') \qquad (6.52)$$

where the I_{pi}s are given coverage intervals on the domain of Z' (using (6.49), these intervals can be easily transformed to the domain of Z). Alternatively, (6.52) allows the calculation of the shortest coverage intervals if the p_is are given.

Cases 1 and 2

It is only for cases 1 and 2 that the conventional procedure described in section 4.6 applies. To the extent that we have assumed that both priors $f(x)$ have the same expectations and variances, the conventional method makes no distinction between these two cases. Thus, the (posterior) best estimate of Z becomes $z_e = x_e + \bar{y}$ and the transformation (6.49) yields $z'_e = 0$. With respect to the standard uncertainty associated with this estimate, recall that in the conventional treatment u_y is taken as equal to σ. Therefore, from the LPU $u_z^2 = u_x^2 + \sigma^2$ and from (6.49) $u_{z'} = 1$, regardless of the ratio η.

The Bayesian treatment yields symmetric pdfs $f_1(z')$ and $f_2(z')$ with respect to zero, thus $z'_e = 0$. To obtain $u_{z'1}$ and $u_{z'2}$ we could integrate (6.51). However, it is more convenient to use the fact that

$$u_z^2 = u_x^2 + u_y^2$$

follows strictly from the LPU for the linear model $Z = X + Y$ with no prior information for Z, whatever the pdfs for X and Y. Equivalently

$$u_{z'i}^2 = \frac{1 + \lambda^2}{1 + \eta^2} \qquad i = 1, 2 \qquad (6.53)$$

where

$$\lambda = \frac{u_y}{u_x}.$$

Now, the standard uncertainty u_x is given—either directly or as $(x_\beta - x_\alpha)/\sqrt{12}$—whereas the Bayesian standard uncertainty u_y has to be derived from the information pertaining to Y. This information is encoded in the joint assessment (6.46). Integrating out the nuisance parameter V we get the posterior $f(y|d) \propto S(y)$, which is a Student's-t pdf with $n - 1$ degrees of freedom on the variable $(y - \bar{y})/\sigma$. Therefore,

$$u_y^2 = \frac{n - 1}{n - 3}\sigma^2$$

and it follows that

$$\eta^2 = \frac{n - 3}{n - 1}\lambda^2. \qquad (6.54)$$

Note that the Bayesian approach to the uncertainty of Y always yields a larger value than the conventional value σ, and that if n increases without limit $\eta \to \lambda$, both parameters approach zero because σ does.

Use of (6.54) allows (6.53) to be plotted as a function of λ. Results are shown in figure 6.26 for $n = 4$ (full curve) and $n = 10$ (broken curve). It is seen that the Bayesian standard uncertainties tend to one (the conventional result) for λ small and increase noticeably as this parameter grows, i.e. as the uncertainty of Y knowing only the data d is augmented relative to the prior uncertainty of X.

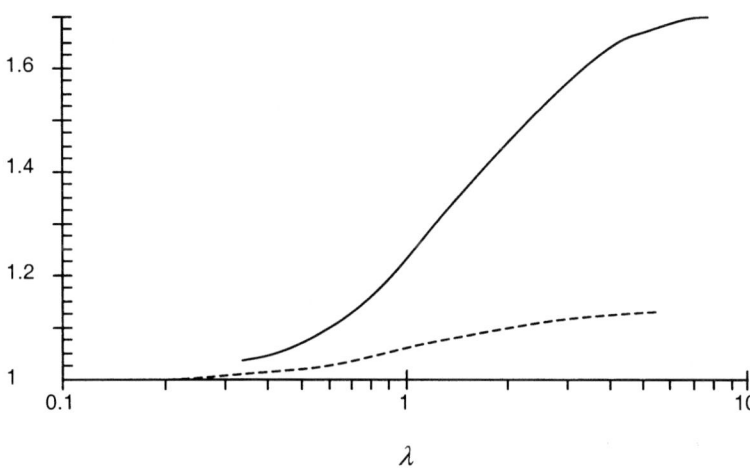

Figure 6.26. Bayesian posterior standard uncertainties for the quantity Z', cases 1 and 2: full curve, $n = 4$; broken curve, $n = 10$. The conventional standard uncertainties are always equal to one, whatever the values of n and λ (or η).

This increase is more pronounced as the number n is reduced, i.e. for given u_x, the less information there is about Y, the greater the uncertainty associated with the estimate of Z' becomes. The upper limit is $u'_z \to \sqrt{3}$ for $n = 4$ and $\lambda \to \infty$.

Let us turn now to the coverage probabilities. By way of example let us choose arbitrarily symmetric coverage intervals with limits $\pm 2u_{z'i}$, $i = 1, 2$. This is equivalent to choosing a coverage factor $k_p = 2$. In the Bayesian treatment (6.52) was integrated numerically over these intervals. In the conventional treatment the confidence level was obtained by numerically integrating an equation similar to (6.52), where the integrand is a t-distribution with v_{eff} degrees of freedom and where the integration interval is $(-2, 2)$. From the WS formula, equation (4.7), and under the assumption that there is 'no uncertainty associated with the uncertainty u_x', the effective degrees of freedom are

$$v_{\text{eff}} = (n-1)\left(\frac{u_z}{\sigma}\right)^4 = (n-1)\left(1 + \frac{n-1}{n-3}\frac{1}{\lambda^2}\right)^2.$$

Results are depicted in figure 6.27 for $n = 4$ (full curves) and $n = 10$ (broken curves). Case 1 is identified by full circles and case 2 by open circles. It is seen that the Bayesian coverage probabilities for given n tend to common values for $\lambda > 1$. As this ratio is decreased, i.e. as the information pertaining to Y diminishes relative to that of X, the pdf $f_1(z')$ retains the normal form while $f_2(z')$ tends to a rectangular density. This causes a definite increase of p_2 over p_1. In fact, for a fixed coverage factor k_p the coverage probability of a rectangular pdf is equal to

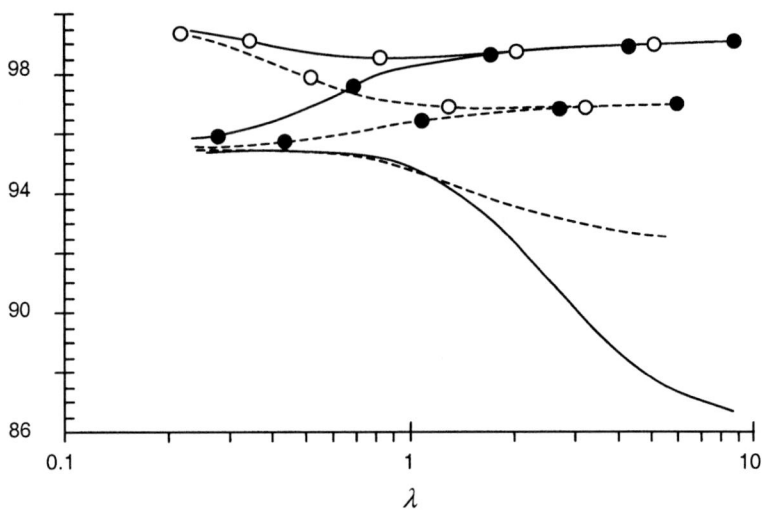

Figure 6.27. Bayesian coverage probabilities for cases 1 (full circles) and 2 (open circles). Conventional confidence levels are depicted as lines without symbols. Full curves, $n = 4$; broken curves, $n = 10$. Coverage intervals in the domain of Z' are $(-2u_{z'}, 2u_{z'})$.

$k_p/\sqrt{3}$. Therefore, taking $k_p = 2$ means that p_2 will eventually attain and even surpass 100 % for very small values of λ. The effect of increasing n for fixed λ is to decrease the coverage probability. This effect is more pronounced as λ grows and is due to the fact that the coverage interval decreases, i.e. the same coverage factor multiplies a smaller standard uncertainty, see figure 6.26.

Lines without symbols in figure 6.27 correspond to the conventional confidence levels. No difference exists between cases 1 and 2 because in the conventional approach the 'effective degrees of freedom' and hence the confidence level depend not on the shape of $f(x)$ but only on its standard uncertainty. According to the *Guide* (Clause 6.3.3), the confidence interval $(-2, 2)$ on the domain of Z' is expected to produce a confidence level of approximately 95 %. It is seen that this happens only for small λ or, equivalently, for small η. For $\lambda > 1$ the confidence levels are notably low *vis-à-vis* the Bayesian coverage probabilities, especially for n small and for case 2. This is recognized in Clause G.6.5 of the *Guide*, where it is stated that the conventional method should not be used if u_x dominates over σ and $f(x)$ is rectangular.

Note that if $n \leq 3$ neither the Bayesian standard uncertainty u_y nor the parameter λ can be defined. However, the coverage probability corresponding to any given coverage interval can be computed from (6.52) even for n as low as 2.

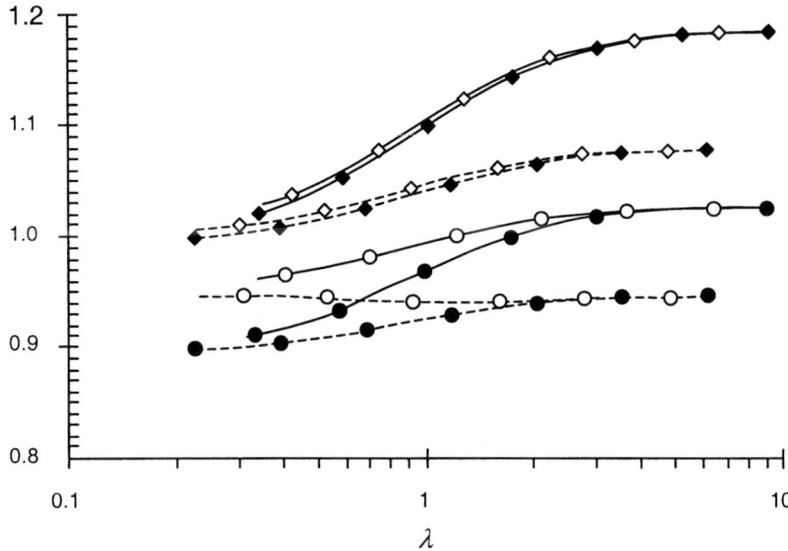

Figure 6.28. Bayesian posterior standard uncertainties for the quantity Z' with $\sigma' = 2$ and $\mu' = 0$: case 3, full circles; case 4, open circles; case 5, full diamonds; case 6, open diamonds; full curves, $n = 4$; broken curves, $n = 10$.

Cases 3 through 6

These cases can be analysed only from the Bayesian viewpoint. Results depend on two additional parameters: the ratio σ' of the standard deviation of the prior density of Z to the characteristic dispersion $(u_x^2 + \sigma^2)^{1/2}$; and the ratio μ' whose numerator is the displacement of the expectation value z_{eb} of the prior density of Z with respect to the value $x_e + \bar{y}$ that is obtained for z_{ea} when there is no prior knowledge about Z. Results are presented in figures 6.28 and 6.29 for $\sigma' = 2$ and $\mu' = 0$. The latter choice determines symmetrical posterior pdfs for Z' about zero, and thus $z'_e = 0$.

Figure 6.28 depicts the standard uncertainties $u_{z'}$; figure 6.29 shows the coverage probabilities p for coverage intervals limited by $\pm 2u_{z'}$. Full curves are for $n = 4$ and broken curves for $n = 10$. It can be seen that the curves corresponding to cases 3 (full circles) and 4 (open circles) merge for large values of λ and separate otherwise. The same happens for cases 5 (full diamonds) and 6 (open diamonds). The reason is that the posterior densities corresponding to these pairs of cases are equal to the products of $f_1(z')$ and $f_2(z')$ multiplied by common functions of z', respectively. As explained in relation to figure 6.27, the difference between $f_1(z')$ and $f_2(z')$ is negligible for large λ. Instead, for small λ, the influence of the rectangular tendency of $f_2(z')$ is transferred to $f_4(z')$ and

218   *Bayesian inference*

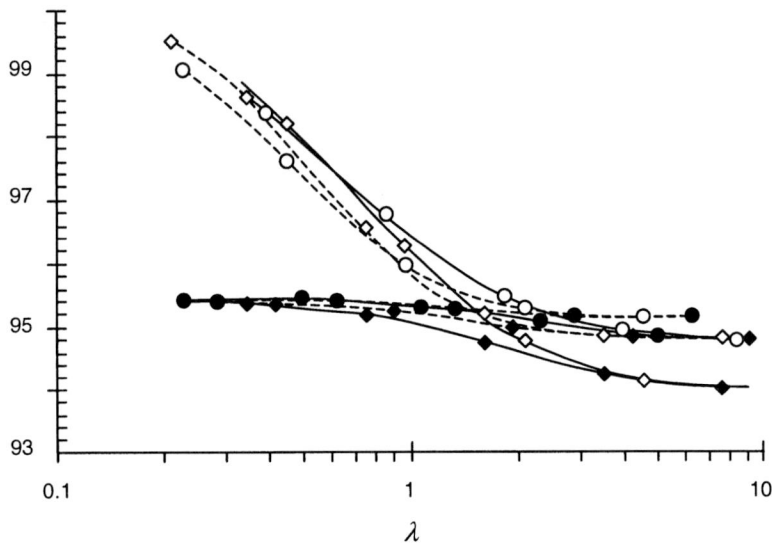

Figure 6.29. Bayesian coverage probabilities for coverage intervals $(-2u_{z'}, 2u_{z'})$ in the domain of Z' with $\sigma' = 2$ and $\mu' = 0$: case 3, full circles; case 4, open circles; case 5, full diamonds; case 6, open diamonds; full curves, $n = 4$; broken curves, $n = 10$.

$f_6(z')$. It is also seen that higher standard uncertainties are obtained using for Z a rectangular prior (diamonds) rather than a normal prior (circles). Finally, *ceteris paribus*, increasing the number of observations of the quantity Y (broken curves) leads to an expected decrease in the standard uncertainties.

The effect of varying σ' is not shown. However, it may readily be seen that, as this parameter is increased, the relative influence of the prior information about Z is diminished. Therefore, the curves corresponding to cases 3 and 5 (full symbols) will tend to those of case 1, while the curves corresponding to cases 4 and 6 (open symbols) will tend to those of case 2.

Finally a word about skewness. The pdfs corresponding to cases 1 and 2 are always symmetric about $z' = 0$. The same is true for the pdfs corresponding to cases 3 through 6 if $\mu' = 0$. However, the latter pdfs are skewed if μ' is allowed to differ from zero. No numerical results are presented, as they would not contribute much to this discussion. But it should be noted that a skewed pdf means, on the one hand, that the corresponding expectation values have to be calculated from (6.50) and, on the other hand, that the shortest coverage interval corresponding to a given coverage probability will be asymmetric with respect to the best estimate. In this case, the concept of expanded uncertainty breaks down and has to be replaced by that of left and right coverage uncertainties.

6.9.4 Known variance of the observations of Y

The model $Z = X + Y$ with normal priors for X and Z and with known variance ϑ of the individual observations of Y was considered in [50]. A more concise derivation of the posterior expectations and variances is given next. Thus, with the likelihood

$$l(y|d) \propto \exp\left[-\frac{1}{2}\frac{\sum(y_i - y)^2}{\vartheta}\right]$$

the posterior joint pdf becomes

$$f(x, y, z) \propto N(y; \bar{y}, \sigma_y) N(x; x_e, u_x) N(z; z_e, u_z) \delta(x + y - z) \quad (6.55)$$

where

$$\sigma_y^2 = \frac{\vartheta}{n}.$$

Integration of (6.55) over y yields

$$f(x, z) \propto N(z - x; \bar{y}, \sigma_y) N(x; x_e, u_x) N(z; z_e, u_z)$$

or, equivalently,

$$f(x, z) \propto \exp\left(-\frac{\phi}{2}\right)$$

where

$$\phi = x^2 \phi_{xx} + z^2 \phi_{zz} - 2x\phi_x - 2z\phi_z - 2xz\phi_{xz} + \text{other terms}$$

$$\phi_{xx} = \frac{1}{\sigma_y^2} + \frac{1}{u_x^2}$$

$$\phi_{zz} = \frac{1}{\sigma_y^2} + \frac{1}{u_z^2}$$

$$\phi_x = \frac{x_e}{u_x^2} - \frac{\bar{y}}{\sigma_y^2}$$

$$\phi_z = \frac{z_e}{u_z^2} + \frac{\bar{y}}{\sigma_y^2}$$

$$\phi_{xz} = \frac{1}{\sigma_y^2}$$

and the 'other terms' do not contain x or z.

Comparing corresponding terms of ϕ with those of the exponent in the bivariate normal pdf

$$f(x, z) \propto \exp\left\{\frac{-1}{2(1-\rho^2)}\left[\frac{(x - x_{ea})^2}{u_{xa}^2} + \frac{(z - z_{ea})^2}{u_{za}^2} - 2\rho\frac{(x - x_{ea})(z - z_{ea})}{u_{xa} u_{ya}}\right]\right\}$$

where
$$\rho = \frac{u_{xa,za}}{u_{xa}u_{za}}$$

the following expressions for the posterior ('after') parameters are obtained:

$$x_{ea} = \left(\frac{z_e - \bar{y}}{u_z^2 + \sigma_y^2} + \frac{x_e}{u_x^2}\right)\left(\frac{1}{u_z^2 + \sigma_y^2} + \frac{1}{u_x^2}\right)^{-1}$$

$$z_{ea} = \left(\frac{x_e + \bar{y}}{u_x^2 + \sigma_y^2} + \frac{z_e}{u_z^2}\right)\left(\frac{1}{u_x^2 + \sigma_y^2} + \frac{1}{u_z^2}\right)^{-1}$$

$$\frac{1}{u_{xa}^2} = \frac{1}{u_z^2 + \sigma_y^2} + \frac{1}{u_x^2}$$

$$\frac{1}{u_{za}^2} = \frac{1}{u_x^2 + \sigma_y^2} + \frac{1}{u_z^2}.$$

The posterior expectation value and variance of Y can also be obtained by integrating (6.55) over x or z and following a similar treatment. The results are

$$y_{ea} = \left(\frac{z_e - x_e}{u_z^2 + u_x^2} + \frac{\bar{y}}{\sigma_y^2}\right)\left(\frac{1}{u_z^2 + u_x^2} + \frac{1}{\sigma_y^2}\right)^{-1}$$

$$\frac{1}{u_{ya}^2} = \frac{1}{u_z^2 + u_x^2} + \frac{1}{\sigma_y^2}.$$

A posteriori all three variables are correlated with covariances (or mutual uncertainties)

$$u_{x,y} = -\frac{u_x^2 \sigma_y^2}{u_x^2 + \sigma_y^2 + u_z^2}$$

$$u_{x,z} = \frac{u_x^2 u_z^2}{u_x^2 + \sigma_y^2 + u_z^2}$$

$$u_{y,z} = \frac{\sigma_y^2 u_z^2}{u_x^2 + \sigma_y^2 + u_z^2}.$$

6.9.5 Summary and conclusions

In this section the Bayesian framework was applied to a very simple measurement model with different types of prior information. The analysis uses the Bayes' theorem and leads to a formal expression for the posterior pdf associated with the measurand. From this pdf, regarded as known, the expectation value, the standard uncertainty and the coverage probability are obtained. Calculations must, in general, be done numerically, although for some situations analytic results can be obtained (as for the standard uncertainties in cases 1 and 2 or in the case considered in section 6.9.4).

Results show that the conventional method may produce confidence levels that differ substantially from the more rigorous coverage probabilities obtained with Bayesian analysis. Moreover, the latter automatically incorporates the LPU (valid only for linear or linearized models) and docs not need to use the WS formula in connection with the doubtful concept of 'effective number of degrees of freedom' based on 'relative uncertainties of the uncertainties'. These reasons all point to Bayesian analysis as being the preferred means with which to evaluate the measurement uncertainty and the related concepts of coverage probability, coverage interval and coverage (or expanded) uncertainty.

Appendix: Probability tables

Table A1. The standard normal cumulative function $\Phi(x)$.

x	0 5	1 6	2 7	3 8	4 9
−3.0	0.0013 0.0002	0.0010 0.0002	0.0007 0.0001	0.0005 0.0001	0.0003 0.0000
−2.9	0.0019 0.0016	0.0018 0.0015	0.0017 0.0015	0.0017 0.0014	0.0016 0.0014
−2.8	0.0026 0.0022	0.0025 0.0021	0.0024 0.0021	0.0023 0.0020	0.0023 0.0019
−2.7	0.0035 0.0030	0.0034 0.0029	0.0033 0.0028	0.0032 0.0027	0.0031 0.0026
−2.6	0.0047 0.0040	0.0045 0.0039	0.0044 0.0038	0.0043 0.0037	0.0041 0.0036
−2.5	0.0062 0.0054	0.0060 0.0052	0.0059 0.0051	0.0057 0.0049	0.0055 0.0048
−2.4	0.0082 0.0071	0.0080 0.0069	0.0078 0.0068	0.0075 0.0066	0.0073 0.0064
−2.3	0.0107 0.0094	0.0104 0.0091	0.0102 0.0089	0.0099 0.0087	0.0096 0.0084
−2.2	0.0139 0.0122	0.0136 0.0119	0.0132 0.0116	0.0129 0.0113	0.0126 0.0110
−2.1	0.0179 0.0158	0.0174 0.0154	0.0170 0.0150	0.0166 0.0146	0.0162 0.0143
−2.0	0.0228 0.0202	0.0222 0.0197	0.0217 0.0192	0.0212 0.0188	0.0207 0.0183
−1.9	0.0287 0.0256	0.0281 0.0250	0.0274 0.0244	0.0268 0.0238	0.0262 0.0233
−1.8	0.0359 0.0322	0.0352 0.0314	0.0344 0.0307	0.0336 0.0300	0.0329 0.0294
−1.7	0.0446 0.0401	0.0436. 0.0392	0.0427 0.0384	0.0418 0.0375	0.0409 0.0367
−1.6	0.0548 0.0495	0.0537 0.0485	0.0526 0.0475	0.0516 0.0465	0.0505 0.0455

Table A1. (Continued)

x	0 / 5	1 / 6	2 / 7	3 / 8	4 / 9
−1.5	0.0668	0.0655	0.0643	0.0630	0.0618
	0.0606	0.0594	0.0582	0.0570	0.0559
−1.4	0.0808	0.0793	0.0778	0.0764	0.0749
	0.0735	0.0722	0.0708	0.0694	0.0681
−1.3	0.0968	0.0951	0.0934	0.0918	0.0901
	0.0885	0.0869	0.0853	0.0838	0.0823
−1.2	0.1151	0.1131	0.1112	0.1093	0.1075
	0.1056	0.1038	0.1020	0.1003	0.0985
−1.1	0.1357	0.1335	0.1314	0.1292	0.1271
	0.1251	0.1230	0.1210	0.1190	0.1170
−1.0	0.1587	0.1562	0.1539	0.1515	0.1492
	0.1469	0.1446	0.1423	0.1401	0.1379
−0.9	0.1841	0.1814	0.1788	0.1762	0.1736
	0.1711	0.1685	0.1660	0.1635	0.1611
−0.8	0.2119	0.2090	0.2061	0.2033	0.2005
	0.1977	0.1949	0.1922	0.1894	0.1867
−0.7	0.2420	0.2389	0.2358	0.2327	0.2297
	0.2266	0.2236	0.2206	0.2177	0.2148
−0.6	0.2743	0.2709	0.2676	0.2643	0.2611
	0.2578	0.2546	0.2514	0.2483	0.2451
−0.5	0.3085	0.3050	0.3015	0.2981	0.2946
	0.2912	0.2877	0.2843	0.2810	0.2776
−0.4	0.3446	0.3409	0.3372	0.3336	0.3300
	0.3264	0.3228	0.3192	0.3156	0.3121
−0.3	0.3821	0.3783	0.3745	0.3707	0.3669
	0.3632	0.3594	0.3557	0.3520	0.3483
−0.2	0.4207	0.4168	0.4129	0.4090	0.4052
	0.4013	0.3974	0.3936	0.3897	0.3859
−0.1	0.4602	0.4562	0.4552	0.4483	0.4443
	0.4404	0.4364	0.4325	0.4286	0.4247
−0.0	0.5000	0.4960	0.4920	0.4880	0.4840
	0.4801	0.4761	0.4721	0.4681	0.4641
0.0	0.5000	0.5040	0.5080	0.5120	0.5160
	0.5199	0.5239	0.5279	0.5319	0.5359
0.1	0.5398	0.5438	0.5478	0.5517	0.5557
	0.5596	0.5636	0.5675	0.5714	0.5753
0.2	0.5793	0.5832	0.5871	0.5910	0.5948
	0.5987	0.6026	0.6064	0.6103	0.6141
0.3	0.6179	0.6217	0.6255	0.6293	0.6331
	0.6368	0.6406	0.6443	0.6480	0.6517
0.4	0.6554	0.6591	0.6628	0.6664	0.6700
	0.6736	0.6772	0.6808	0.6844	0.6879
0.5	0.6915	0.6950	0.6985	0.7019	0.7054

Table A1. (Continued)

x	0 5	1 6	2 7	3 8	4 9
0.6	0.7088 0.7257 0.7422	0.7123 0.7291 0.7454	0.7157 0.7324 0.7486	0.7190 0.7357 0.7517	0.7224 0.7389 0.7549
0.7	0.7580 0.7734	0.7611 0.7764	0.7642 0.7794	0.7673 0.7823	0.7703 0.7852
0.8	0.7881 0.8023	0.7910 0.8051	0.7939 0.8078	0.7967 0.8106	0.7995 0.8133
0.9	0.8159 0.8289	0.8166 0.8315	0.8212 0.8340	0.8238 0.8365	0.8264 0.8389
1.0	0.8413 0.8531	0.8438 0.8554	0.8461 0.8577	0.8485 0.8599	0.8508 0.8621
1.1	0.8643 0.8749	0.8665 0.8770	0.8686 0.8790	0.8708 0.8810	0.8729 0.8830
1.2	0.8849 0.8944	0.8869 0.8962	0.8888 0.8980	0.8907 0.8997	0.8925 0.9015
1.3	0.9032 0.9115	0.9049 0.9131	0.9066 0.9147	0.9082 0.9162	0.9099 0.9177
1.4	0.9192 0.9265	0.9207 0.9278	0.9222 0.9292	0.9236 0.9306	0.9251 0.9319
1.5	0.9332 0.9394	0.9345 0.9406	0.9357 0.9418	0.9370 0.9430	0.9382 0.9441
1.6	0.9452 0.9505	0.9463 0.9515	0.9474 0.9525	0.9484 0.9535	0.9495 0.9545
1.7	0.9554 0.9599	0.9564 0.9608	0.9573 0.9616	0.9582 0.9625	0.9591 0.9633
1.8	0.9641 0.9678	0.9648 0.9686	0.9656 0.9693	0.9664 0.9700	0.9671<.br>0.9706
1.9	0.9713 0.9744	0.9719 0.9750	0.9726 0.9756	0.9732 0.9762	0.9738 0.9767
2.0	0.9772 0.9798	0.9778 0.9803	0.9783 0.9808	0.9788 0.9812	0.9793 0.9817
2.1	0.9821 0.9842	0.9826 0.9846	0.9830 0.9850	0.9834 0.9854	0.9838 0.9857
2.2	0.9861 0.9878	0.9864 0.9881	0.9868 0.9884	0.9871 0.9887	0.9874 0.9890
2.3	0.9893 0.9906	0.9896 0.9909	0.9898 0.9911	0.9901 0.9913	0.9904 0.9916
2.4	0.9918 0.9929	0.9920 0.9931	0.9922 0.9932	0.9925 0.9934	0.9927 0.9936
2.5	0.9938 0.9946	0.9940 0.9948	0.9941 0.9949	0.9943 0.9951	0.9945 0.9952
2.6	0.9953 0.9960	0.9955 0.9961	0.9956 0.9962	0.9957 0.9963	0.9959 0.9964

Table A1. (Continued)

x	0 5	1 6	2 7	3 8	4 9
2.7	0.9965	0.9966	0.9967	0.9968	0.9969
	0.9970	0.9971	0.9972	0.9973	0.9974
2.8	0.9974	0.9975	0.9976	0.9977	0.9977
	0.9978	0.9979	0.9979	0.9980	0.9981
2.9	0.9981	0.9982	0.9982	0.9983	0.9984
	0.9984	0.9985	0.9985	0.9986	0.9986
3.0	0.9987	0.9990	0.9993	0.9995	0.9997
	0.9998	0.9998	0.9999	0.9999	1.0000

Table A2. Student's t-quantiles $t_{p,\nu}$.

ν	\multicolumn{6}{c}{Coverage probability p}
	0.6827	0.9000	0.9500	0.9545	0.9900	0.9973
1	1.84	6.31	12.71	13.97	63.66	235.80
2	1.32	2.92	4.30	4.53	9.92	19.21
3	1.20	2.35	3.18	3.31	5.84	9.22
4	1.14	2.13	2.78	2.87	4.60	6.62
5	1.11	2.02	2.57	2.65	4.03	5.51
6	1.09	1.94	2.45	2.52	3.71	4.90
7	1.08	1.89	2.36	2.43	3.50	4.53
8	1.07	1.86	2.31	2.37	3.36	4.28
9	1.06	1.83	2.26	2.32	3.25	4.09
10	1.05	1.81	2.23	2.28	3.17	3.96
11	1.05	1.80	2.20	2.25	3.11	3.85
12	1.04	1.78	2.18	2.23	3.05	3.76
13	1.04	1.77	2.16	2.21	3.01	3.69
14	1.04	1.76	2.14	2.20	2.98	3.64
15	1.03	1.75	2.13	2.18	2.95	3.59
16	1.03	1.75	2.12	2.17	2.92	3.54
17	1.03	1.74	2.11	2.16	2.90	3.51
18	1.03	1.73	2.10	2.15	2.88	3.48
19	1.03	1.73	2.09	2.14	2.86	3.45
20	1.03	1.72	2.09	2.13	2.85	3.42
25	1.02	1.71	2.06	2.11	2.79	3.33
30	1.02	1.70	2.04	2.09	2.75	3.27
35	1.01	1.70	2.03	2.07	2.72	3.23
40	1.01	1.68	2.02	2.06	2.70	3.20

Appendix: Probability tables

Table A2. (Continued)

	\multicolumn{6}{c}{Coverage probability p}
v	0.6827	0.9000	0.9500	0.9545	0.9900	0.9973
45	1.01	1.68	2.01	2.06	2.69	3.18
50	1.01	1.68	2.01	2.05	2.68	3.16
100	1.005	1.660	1.984	2.025	2.626	3.077
∞	1.000	1.645	1.960	2.000	2.576	3.000

Table A3. The consumer's risk for inspection of workpieces.

τ	κ	\multicolumn{4}{c}{ρ}
		1	2	3	4
1	0.8	20.9	10.2	2.7	0.4
	0.9	23.5	11.6	3.0	0.5
	1.0	26.0	12.8	3.2	0.5
	1.1	28.5	13.8	3.3	0.5
	1.2	30.9	14.5	3.4	0.5
2	0.8	7.8	1.9	0.2	0.0
	0.9	9.9	3.0	0.4	0.0
	1.0	12.3	4.1	0.6	0.0
	1.1	14.8	5.1	0.7	0.0
	1.2	17.3	5.9	0.7	0.0
3	0.8	3.7	0.5	0.0	0.0
	0.9	5.5	1.2	0.1	0.0
	1.0	7.7	2.2	0.2	0.0
	1.1	10.3	3.3	0.4	0.0
	1.2	13.0	4.2	0.4	0.0
4	0.8	1.9	0.2	0.0	0.0
	0.9	3.4	0.6	0.0	0.0
	1.0	5.6	1.5	0.2	0.0
	1.1	8.2	2.6	0.3	0.0
	1.2	11.2	3.6	0.3	0.0
5	0.8	1.0	0.0	0.0	0.0
	0.9	2.3	0.3	0.0	0.0
	1.0	4.3	1.1	0.1	0.0
	1.1	7.1	2.3	0.2	0.0
	1.2	10.2	3.4	0.3	0.0

Table A4. The producer's risk for inspection of workpieces.

		\multicolumn{4}{c}{ρ}
τ	κ	1	2	3	4
1	0.8	15.3	5.4	1.0	0.1
	0.9	12.3	3.1	0.3	0.0
	1.0	9.8	1.7	0.1	0.0
	1.1	7.7	0.8	0.0	0.0
	1.2	6.0	0.4	0.0	0.0
2	0.8	13.1	5.5	1.1	0.1
	0.9	9.6	2.8	0.4	0.0
	1.0	6.9	1.2	0.1	0.0
	1.1	4.8	0.5	0.0	0.0
	1.2	3.2	0.2	0.0	0.0
3	0.8	11.7	5.7	1.2	0.1
	0.9	8.0	2.6	0.4	0.0
	1.0	5.2	1.0	0.1	0.0
	1.1	3.1	0.3	0.0	0.0
	1.2	1.8	0.1	0.0	0.0
4	0.8	11.1	5.9	1.3	0.1
	0.9	7.0	2.5	0.4	0.0
	1.0	4.1	0.8	0.1	0.0
	1.1	2.1	0.2	0.0	0.0
	1.2	1.0	0.0	0.0	0.0
5	0.8	10.7	6.0	1.3	0.1
	0.9	6.4	2.5	0.4	0.0
	1.0	3.4	0.7	0.1	0.0
	1.1	1.5	0.1	0.0	0.0
	1.2	0.6	0.0	0.0	0.0

Table A5. The consumer's risk for instrument verification.

		\multicolumn{5}{c}{$\tau\rho$}
τ	$\delta/(\tau\rho)$	1	2	3	4	5
1	0.0	48.0	15.7	3.4	0.5	0.0
	0.2	48.4	17.4	5.0	1.2	0.2
	0.4	49.7	22.2	10.3	4.5	1.7
	0.6	51.8	29.8	19.8	12.9	7.9
	0.8	54.5	39.4	33.6	28.6	24.0
	1.0	57.9	50.2	50.0	50.0	50.0

Table A5. (Continued)

τ	$\delta/(\tau\rho)$	\multicolumn{5}{c}{$\tau\rho$}
		1	2	3	4	5
$\frac{1}{2}$	0.0	37.1	7.4	0.7	0.0	0.0
	0.2	37.9	9.2	1.7	0.2	0.0
	0.4	40.1	14.8	5.4	1.6	0.4
	0.6	43.6	23.9	14.2	7.6	3.7
	0.8	48.3	36.1	29.6	23.7	18.6
	1.0	53.7	50.0	50.0	50.0	50.0
$\frac{1}{3}$	0.0	34.3	5.8	0.4	0.0	0.0
	0.2	35.1	7.6	1.2	0.1	0.0
	0.4	37.7	13.1	4.4	1.1	0.2
	0.6	41.7	22.5	12.7	6.5	2.9
	0.8	46.9	35.2	28.5	22.4	17.1
	1.0	52.9	50.0	50.0	50.0	50.0
$\frac{1}{4}$	0.0	33.2	5.2	0.4	0.0	0.0
	0.2	34.1	7.0	1.0	0.1	0.0
	0.4	36.7	12.5	4.0	1.0	0.2
	0.6	40.9	22.0	12.2	6.0	2.6
	0.8	46.3	34.9	28.0	21.9	16.6
	1.0	52.6	50.0	50.0	50.0	50.0
$\frac{1}{5}$	0.0	32.7	5.0	0.3	0.0	0.0
	0.2	33.6	6.8	1.0	0.1	0.0
	0.4	36.3	12.3	3.9	0.9	0.2
	0.6	40.6	21.7	12.0	5.8	2.5
	0.8	46.1	34.8	27.8	21.6	16.3
	1.0	52.5	50.0	50.0	50.0	50.0

Table A6. The producer's risk for instrument verification.

τ	$\delta/(\tau\rho)$	\multicolumn{5}{c}{$\tau\rho$}
		1	2	3	4	5
1	0.2	51.6	82.6	95.0	98.8	99.8
	0.4	50.3	77.8	89.7	95.5	98.3
	0.6	48.2	70.2	80.2	87.1	92.1
	0.8	45.5	60.6	66.4	71.4	76.0
	1.0	42.1	49.8	50.0	50.0	50.0
	2.0	22.3	7.9	1.7	0.2	0.0
	3.0	7.6	0.2	0.0	0.0	0.0

Table A6. (Continued)

τ	$\delta/(\tau\rho)$	\multicolumn{5}{c}{$\tau\rho$}
		1	2	3	4	5
$\frac{1}{2}$	0.2	62.1	90.8	98.3	99.8	100
	0.4	59.9	85.2	94.6	98.4	99.6
	0.6	56.4	76.1	85.8	92.4	96.3
	0.8	51.7	63.9	70.4	76.3	81.4
	1.0	46.3	50.0	50.0	50.0	50.0
	2.0	18.2	3.7	0.4	0.0	0.0
	3.0	3.7	0.0	0.0	0.0	0.0
$\frac{1}{3}$	0.2	64.9	92.4	98.8	99.9	100
	0.4	62.3	86.9	95.6	98.9	99.8
	0.6	58.3	77.5	87.3	93.5	97.1
	0.8	53.1	64.8	71.5	77.6	82.9
	1.0	47.1	50.0	50.0	50.0	50.0
	2.0	16.9	2.9	0.2	0.0	0.0
	3.0	2.9	0.0	0.0	0.0	0.0
$\frac{1}{4}$	0.2	65.9	93.0	99.0	99.9	100
	0.4	63.3	87.5	96.0	99.0	99.8
	0.6	59.1	78.0	87.8	94.0	97.4
	0.8	53.7	65.1	72.0	78.1	83.4
	1.0	47.4	50.0	50.0	50.0	50.0
	2.0	16.4	2.6	0.2	0.0	0.0
	3.0	2.6	0.0	0.0	0.0	0.0
$\frac{1}{5}$	0.2	66.4	93.2	99.0	99.9	100
	0.4	63.7	87.7	96.1	99.1	99.8
	0.6	59.4	78.3	88.0	94.2	97.5
	0.8	53.9	65.2	72.2	78.4	83.7
	1.0	47.5	50.0	50.0	50.0	50.0
	2.0	16.2	2.5	0.2	0.0	0.0
	3.0	2.5	0.0	0.0	0.0	0.0

Glossary

Abbreviations

BIPM	Bureau International des Poids et Mesures
BT	Bayes' theorem
CLT	Central limit theorem
DMM	Digital multimeter
emf	Electromotive force
GLPU	Generalized law of propagation of uncertainties
Guide	ISO Guide to the Expression of Uncertainty in Measurement (also GUM)
ISO	International Organization for Standardization
ITS	International Temperature Scale
LPU	Law of propagation of uncertainties
LSL	Lower specification limit
MFC	Multifunction calibrator
pdf	Probability density function
PME	Principle of maximum entropy
SPRT	Standard Platinum Resistance Thermometer
TA	Type A uncertainty evaluation procedure
TUR	Test uncertainty ratio
USL	Upper specification limit
VIM	ISO International Vocabulary of Basic and General Terms in Metrology
WS	Welch–Satterthwaite (formula)

Symbols

The following list is ordered according to the topic or context in which each symbol appears. Many symbols have different meanings, they may appear more than once.

Quantities and pdfs

$Q, Z, \ldots$	General quantities interpreted as random variables
$q, z, \ldots$	Possible values of the quantities
$f(q)$, $f_Q(q)$	(Marginal) pdf, frequency distribution
	(the subscript is used only if needed to avoid confusion)
$f(q, z)$	Joint pdf for the quantities Q and Z
$f(q\|z)$	Conditional pdf of Q given a value of Z
$f(q\|d)$	Posterior pdf of Q given data d
$l(q\|d)$	Likelihood of the data d
D_Q	Domain of the pdf $f(q)$
$E(Q)$	Expectation
$V(Q)$	Variance
$S(Q)$	Standard deviation
$C(Q, Z)$	Covariance between Q and Z
μ	Expectation of a pdf
σ^2	Variance of a pdf, estimate of variance
σ	Standard deviation of a pdf, value of a standard uncertainty
q_e	(Best) estimated value of Q
$\dot{q}_e$	Previous best estimate
$\ddot{q}_e$	Posterior best estimate
u_q	Standard uncertainty associated with the (best) estimate
$u(Q)$	Same as above if Q is a particular quantity
u_q^*	Standard uncertainty associated with the estimate q^*
$\dot{u}_q$	Previous standard uncertainty
$\ddot{u}_q$	Posterior standard uncertainty
u_{rq}	Relative standard uncertainty
$u_{q,z}$	Mutual uncertainty between Q and Z
$r_{q,z}$	Correlation coefficient between Q and Z
ξ	Ratio of relative uncertainties

Special pdfs and functions

$N(x)$	Normal (or Gaussian) pdf
$N_s(x)$	Standard normal pdf
$\Phi(x)$	Standard normal cumulative function
h, ϕ, δ, e	Parameters of the trapezoidal pdf
$R(x)$	Rectangular pdf
W	Width of a rectangular pdf
E	Quantity for the logarithmic pdf
e_e, ε	Parameters of the logarithmic pdf
$S_\nu(t)$	Student's-t pdf

232   Glossary

T	Quantity whose pdf is Student's-t
K_ν	Normalization constant of Student's-t pdf
ν	Degrees of freedom of Student's-t pdf
$t_{p,\nu}$	Quantile of Student's-t pdf
r^2	Variance of Student's-t pdf
a, b, c	Parameters of the 'distorted U' pdf
$\tilde{q}, \delta$	Parameters of the 'cusp' pdf
S	Entropy of a pdf
A, λ	Parameters of a maximun entropy pdf
$\exp(x)$	Exponential function
$H(x)$	Heaviside unit step function
$\Gamma(x)$	Gamma function
$\delta(x)$	Dirac's delta function

Probabilities, expanded uncertainties

$A, B, \ldots$	Events
$P(A)$	Probability of event A occurring
p	Probability, coverage probability, proportion
S	Specification interval
μ	Nominal value of the specification interval
ε	Half the tolerance of the specification interval
A_p, A	Acceptance interval
I_p	Coverage interval
k_p	Coverage factor
κ_p, κ	Guardband factor
U_{pq}	Expanded uncertainty with coverage probability p

Models

$\mathcal{M}(Z)$	Model function of input quantity Z
$\mathcal{M}$	Model function evaluated at the best estimate
$\mathcal{M}'$	(First) derivative of the model evaluated at the best estimate
$\mathcal{M}''$	Second derivative of the model evaluated at the best estimate
Q_L	Linearized quantity
c_z	Sensitivity coefficient of input quantity Z

Statistics

Θ	Random sequence of equally distributed random variables
θ	Random sample
n	Size of the random sample

$\overline{Q}$	Sample mean
$\overline{q}$	Value of the sample mean
S^2	Sample variance
s^2	Value of the sample variance
s'^2, s_p^2	Pooled estimate of variance
s	Precision of the random sample

Matrix analysis

A	Vector or matrix
$\mathbf{A}^T$	The transpose of **A**
$\mathbf{A}^{-1}$	The inverse of **A**
I	The identity matrix
$\mathbf{S_q}$	Sensitivity matrix
$\mathbf{u_q^2}$	Uncertainty matrix
1	Matrix (or vector) with all elements equal to one
0	Matrix (or vector) with all elements equal to zero
$n_\mathcal{M}$	Number of model relations
n_Q	Number of quantities in vector **Q**

Symbols for special or particular quantities

A	Area, parameter of a calibration curve
a, e, ℓ, l	Lengths
C	Correction
D, d	Difference between two particular quantities
d	Data
E	Modulus of elasticity, energy flux, electromotive force
F	Correction factor
G	Gauged or calculated indication, gravitational constant
g	Acceleration of gravity
h	Enthalpy
I	Indications during calibration, electric current
L	Load
L	Vector of Lagrange multipliers
m	Mass, mass flow
P	Pointer position
P	Vector of primary quantities
p	Pressure
R	Resolution quantity, ratio of quantities
S	(Measurement) standard, Seebeck coefficient
T	Temperature, test quantity (calibrand)

234   *Glossary*

V	Volume, emf, potential difference
X	Resistance
Y	Reactance
α	Coefficient of thermal expansion
γ	Capacitance
ε	Strain
θ	Time interval
λ	Inductance
ρ	Density, resistance
σ	Stress
τ	Measurement conditions
ϕ	Phase shift angle
ω	Frequency

Symbols for special parameters or constants

Bi	Birge ratio
e_n	Normalized error
K	Normalization constant
R	Ideal gas constant, risk, reliability
V, v, ϑ	Variance
δ	Digital step
ε	Maximum permissible error, accuracy
η, λ	Ratio of standard uncertainties
ν, δ, τ, ρ	Parameters to evaluate the producer's and consumer's risks
ϕ	Nonlinearity
χ^2	Sum of squares

Special subscripts

a	Air, adjusted, after
b	Before
C	Consumer
c	Classical, calibration, critical
e	Estimate
eff	Effective
f	Flask
I	Indicated
L	Linear
l	lower
lim	Limit
m	Modified

max	Maximum
min	Minimum
n	Nominal
o	Reference, degree (superscript)
P	Producer
p	Pool
q	Quantities
r	Reference, relative
S, s	(Measurement) standard
t	Thermal expansion, true value
T	Test
u	User, upper
w	Water

General symbols

$a, b, \ldots$	Constants
e, ε	Error bound, half-width of interval
$f(x)$	Function
$f'(x)$	Derivative of $f(x)$
$i, j, k, \ldots$	Integers, indexes
$I, J, \ldots$	Ranges of indexes $i, j, \ldots$
N, n	Total number (of quantities, of random numbers, etc)
r	Ratio
α	Angle
$\propto$	Proportional to
Δ	Used to indicate a difference

References

[1] ANSI/NCSL Z540-1 1994 *Calibration Laboratories and Measuring and Test Equipment—General Requirements* (Boulder, CO: National Conference of Standards Laboratories)

[2] Ballico M 2000 Limitations of the Welch–Satterthwaite approximation for measurement uncertainty calculations *Metrologia* **37** 61–4

[3] Bernardo J and Smith A 2001 *Bayesian Theory* (Chichester: Wiley)

[4] Box G and Tiao G 1992 *Bayesian Inference in Statistical Analysis* (New York: Wiley)

[5] Chiba S and Smith D 1991 A suggested procedure for resolving an anomaly in least-squares data analysis known as 'Peelle's Pertinent Puzzle' and the general implications for nuclear data evaluation *Argonne National Laboratory* ANL/NDM-121

[6] Cox M and Harris P 1999 Up a GUM tree? *The Standard* (Newsletter of the Measurement Quality Division, American Society for Quality) **99** 36–8

[7] Cox M, Daiton M, Harris P and Ridler N 1999 The evaluation of uncertainties in the analysis of calibration data *Proc. 16th IEEE Instrum. and Meas. Technol. Conf.* (Piscataway, NJ: Institute of Electrical and Electronic Engineers)

[8] Cox M, Daiton M and Harris P 2001 Software specifications for uncertainty evaluation and associated statistical analysis *Technical Report* CMSC 10/01 (Teddington, UK: National Physical Laboratory)

[9] DIN 1319-4 1999 *Fundamentals of Metrology—Part 4: Evaluation of Measurement Uncertainty* (in German) (Berlin: Beuth)

[10] EA-4/02 1999 *Expression of the Uncertainty of Measurement in Calibration* (Utrecht: European Cooperation for Accreditation)

[11] Eagle A 1954 A method for handling errors in testing and measuring *Industrial Quality Control* March 10, 10–15

[12] Elster C 2000 Evaluation of measurement uncertainty in the presence of combined random and analogue-to-digital conversion errors *Meas. Sci. Technol.* **11** 1359–63

[13] Estler T 1999 Measurement as inference: fundamental ideas *Ann. CIRP* **48**(2) 611–32

[14] Gill P, Murray W and Wright M 1997 *Practical Optimization* (London: Academic)

[15] Gläser M 2000 Cycles of comparison measurements, uncertainties and efficiencies *Meas. Sci. Technol.* **11** 20–4

[16] Hall B 1999 Uncertainty calculations using object-oriented computer modelling *Meas. Sci. Technol.* **10** 380–6

[17] Hall B 2000 Computer modelling of uncertainty calculations with finite degrees of freedom *Meas. Sci. Technol.* **11** 1335–41

[18] Hall B and Willink R 2001 Does 'Welch–Satterthwaite' make a good uncertainty estimate? *Metrologia* **38** 9–15
[19] Hildebrand F 1976 *Advanced Calculus for Applications* (Englewood Cliffs, NJ: Prentice-Hall)
[20] ISO 1993 *International Vocabulary of Basic and General Terms in Metrology* (Geneva: International Organization for Standardization)
[21] ISO 1993 *Guide to the Expression of Uncertainty in Measurement* corrected and reprinted 1995 (Geneva: International Organization for Standardization)
[22] ISO 10012-1 1992 *Quality Assurance Requirements for Measuring Equipment. Part 1: Metrological Confirmation System for Measuring Equipment* (Geneva: International Organization for Standardization)
[23] ISO 14253-1 1998 *Geometrical Product Specification (GPS)—Inspection by Measurement of Workpieces and Measuring Instruments—Part 1: Decision Rules for Proving Conformance or Non-conformance with Specifications* (Geneva: International Organization for Standardization)
[24] ISO/IEC/EN 17025 1999 *General Requirements for the Competence of Calibration and Testing Laboratories* formerly ISO Guide 25 and EN 45001 (Geneva: International Organization for Standardization)
[25] Jaynes E 1968 Prior probabilities *IEEE Trans. Sys. Sci. Cyber.* **SSC-4** 227–41
[26] Lee P 1989 *Bayesian Statistics: An Introduction* (New York: Oxford University Press)
[27] Lira I and Wöger W 1997 The evaluation of standard uncertainty in the presence of limited resolution of indicating devices *Meas. Sci. Technol.* **8** 441–3
[28] Lira I and Wöger W 1998 The evaluation of the uncertainty associated with a measurement result not corrected for systematic effects *Meas. Sci. Technol.* **9** 1010–11
[29] Lira I and Wöger W 1998 The evaluation of the uncertainty in knowing a directly measured quantity *Meas. Sci. Technol.* **9** 1167–73
[30] Lira I, Camarano D, Paredes Villalobos J and Santiago F 1999 Expression of the uncertainty of measurement in the calibration of thermometers. Part I: Standard platinum resistance thermometers *Metrologia* **36** 107–11
[31] Lira I and Santos P 1999 Expression of the uncertainty of measurement in the calibration of thermometers. Part II: Thermocouples *Metrologia* **36** 415–19
[32] Lira I and Kyriazis G 1999 Bayesian inference from measurement information *Metrologia* **36** 163–9
[33] Lira I 1999 A Bayesian approach to the consumer's and producer's risks in measurement *Metrologia* **36** 397–402
[34] Lira I 2000 Curve adjustment by the least-squares method *Metrologia* **37** 677–81
[35] Lira I 2001 Evaluation of comparison measurements by a least-squares method *Meas. Sci. Technol.* **12** 1167–71
[36] Lira I and Wöger W 2001 Bayesian evaluation of standard uncertainty and coverage probability in a simple measurement model *Meas. Sci. Technol.* **12** 1172–9
[37] Mathioulakis E and Belessiotis V 2000 Uncertainty and traceability in calibration by comparison *Meas. Sci. Technol.* **11** 771–5
[38] Meyer P 1970 *Introductory Probability and Statistical Applications* (Reading, MA: Addison-Wesley)
[39] Mills I 1995 Unity as a unit *Metrologia* **31** 537–41

References

[40] NAMAS NIS 3003 1995 *The Expression of Uncertainty and Confidence in Measurement for Calibration* (Middlesex: NAMAS Executive, NPL)

[41] Olesko K 1994 The measuring of precision: the exact sensibility in early nineteenth-century Germany *The Values of Precision* ed N Wise (Princeton, NJ: Princeton University Press)

[42] Orear J 1982 Least squares when both variables have uncertainties *Am. J. Phys.* **50** 912–16

[43] Phillips S, Estler W, Levenson M and Eberhardt K 1998 Calculation of measurement uncertainty using prior information *J. Res. Natl Inst. Stand. Technol.* **103** 625–32

[44] Preston-Thomas H 1990 The International Temperature Scale of 1990 (ITS-90) *Metrologia* **27** 3–10

[45] Radhakrishna C 1973 *Linear Statistical Inference and its Applications* (New York: Wiley)

[46] Scarpa F 1998 Evaluation of reliable confidence bounds for sensor calibration curves *Metrologia* **35** 1–5

[47] Shannon C and Weaver W 1949 The mathematical theory of communication *Inform. Systems* **8** 51–62

[48] Sutton C and Clarkson M 1993/94 A general approach to comparisons in the presence of drift *Metrologia* **30** 487–93

[49] Taylor B and Kuyatt C 1993 *Guidelines for Evaluating and Expressing the Uncertainty of NIST Measurement Results* NIST TN 1297 (Gaithersburg, MD: NIST)

[50] Tuninsky V and Wöger W 1997 A Bayesian approach to recalibration *Metrologia* **34** 459–65

[51] Weise K 1986 Treatment of uncertainties in precision measurements *IEEE Trans. Instrum. Meas.* **36** 642–5

[52] Weise K and Wöger W 1993 A Bayesian theory of measurement uncertainty *Meas. Sci. Technol.* **4** 1–11

[53] Weise K and Wöger W 1999 *Uncertainty and Measurement Data Evaluation* (in German) (Berlin: Wiley–VCH)

[54] Weise K and Wöger W 2000 Removing modeland data non-conformity in measurement evaluation *Meas. Sci. Technol.* **11** 1649–58

[55] Wöger W 1987 Probability assignment to systematic deviations by the principle of maximum entropy *IEEE Trans. Instrum. Meas.* **36** 655–8

Index to chapters 2–6

Accuracy, 38, 99
Adjustment, 98–101, 160
Analogue display devices, 87

Balance (mass), 89–90, 94, 160–165
Bayes's prior, 174
Bayes' theorem (BT)
   in terms of pdfs, 170
   in terms of propositions, 167
Binomial distribution, 56
Birge ratio, 144–145, 147, 158
Bridge (electrical), 95
Buoyancy, 90, 155, 157

Calibration
   certificate, 30, 96, 116, 118, 154
   curve, 96–98, 151–152, 160, 164
   general concepts, 29–31, 92–98
   laboratory, 93, 152, 198
Capacitor, 147
Central limit theorem (CLT), 114
Coefficient
   correlation, 53
   drift, 156
   of an adjusted curve, 138
   Seebeck, 154
   sensitivity, 63
   thermal expansion, 33, 89
Comparator, 31, 94, 155, 158
Comparison measurements, 85, 94–95, 131–133, 155–160
Confidence level, 53, 112–113
Confirmation, 98

Conformity assessment, 25, 102
Correction, 30, 33, 40, 82, 88, 96, 99
Covariance, 52
Coverage
   factor, 53, 107, 193
   interval, 53, 107, 192
   probability, 53, 103, 109, 192
   uncertainty, 107–108
Current, electrical, 127

Degrees of freedom, 85, 105, 144
Deviation, 96
Digital display devices, 85–87, 182
Dirac's delta function
   as model prior, 201
   definition, 70
Dominance, 172, 173, 175
Drift, 82, 156

Effective degrees of freedom, 115
Elasticity modulus, 39–43, 65, 77, 117
Energy flux, 79–81
Enthalpy, 79–81
Error
   definition, 36
   maximum permissible, 99
   normalized, 144
   random, 36, 56
   systematic, 36, 88
Estimate
   best, 51
   of expectation, 83

of the uncertainty, 43, 84
of variance, 83
Event, 47
Expectation, 50

Factor
   correction, 33, 40, 82, 88–92, 96–97, 99
   coverage, 53, 103, 108, 193
   guardband, 110, 193
   repeat, 132, 155
   weight, 132, 159
Fixed temperature points, 93, 151
Frequency
   based statistics, 82
   distribution, 49, 75, 82, 111, 117, 175
   of an applied voltage, 147
Frequentialization, 116
Fundamental constants, 149

Gain, 100
Galvanometer, 95
Gamma function, 106
Gauge blocks, 31, 93
Gauging, 93
Generalized law of propagation of uncertainties (GLPU), 124
Gravitational constant, 150
Gravity, acceleration of, 90
Guardbanding, 110–111

Heaviside function, 201

Ice point temperature, 150, 154
Ideal
   gases, 91
   instrument, 86, 97
Inductance, 147
Information theory, 188
International temperature scale (ITS), 151
Interval
   acceptance, 34, 110–111, 193
   confidence, 112–113

coverage, 53, 107, 192
specification, 26, 34, 109–111, 193

Jeffreys' prior, 174
Jury principle, 173, 188

Lagrange multipliers, 135
Law of propagation of uncertainties (LPU), 64
Least-squares adjustment
   classical, 134, 141
   consistency analysis, 144–148
   in curve fitting, 138–143
   in straight-line fitting, 141–142
   under uncertainty, 135–138
Level of confidence, 53
Levenberg–Marquardt algorithms, 143
Likelihood, 170

Mass
   density, 89–92
   flow rate, 80–81
   scale, 160–163
Material measure, 30, 93
Matrix
   addition, 121
   block, 121
   column, 120
   comparison design, 155
   covariance, 123
   definition, 120
   diagonal, 120
   dimension, 120
   evaluation, 157
   identity, 120
   inverse, 122
   multiplication, 121
   order, 157
   row, 121
   sensitivity, 123
   singular, 122
   square, 120

stacked, 121
summation of elements, 122
symmetric, 120
transpose, 120
uncertainty, 123, 127
unit, 120
juxtaposed, 121
Measurand
  definition, 28, 39
  incomplete definition, 33
Measurement
  comparison, 85, 94–95, 131–133, 155–160
  definition, 28
  direct, 39
  equal variance, 147, 150, 157
  error, 35
  indirect, 39
  intercomparison, 42, 149, 184
  model, 39–43
  redundant, 132, 155
  standard, 30, 92–101, 131, 155
Measuring instrument or system, 29, 81, 92, 94
Metre, 28
Model
  additive, 64
  linear, 64, 125–126, 137–138
  linearized, 63, 66, 124, 135, 202, 207
  measurement, 39–43
  multiplicative, 64
  non-analytic, 65–66, 77
  nonlinear, 66–75
  prior pdf, 201
  probability, 169, 175
Monte Carlo method, 75–76
Multimeter, 142, 200

Newton–Raphson iteration, 124, 136, 140, 143
Non-informative priors, 173–175
Nonlinearity, 99
Normalization condition, 49, 169

Null condition, 95
Numerical instabilities, 147, 164

Oscilloscope, 147
Outliers, 144, 150

Parameter
  location, 175, 188
  nuisance, 176
  of a calibration curve, 98
  of a frequency distribution, 112, 175
  of a pdf, 49–51, 53
  scale, 175
Parent distribution, 175
Peele's puzzle, 145–146
Phase shift, 127, 147
Pooled estimate of variance, 85
Population, 82, 110, 195, 196
Potential difference (voltage, electromotive force), 127, 147, 150, 200
Precision
  definition, 38
  of the random sample, 84
Pressure, 79–81, 88–92
Principle of maximum entropy (PME), 187–192
Probability
  coverage, 53, 103, 109, 192
  function, 47
  interpretation, 49
  model, 169, 175
  of propositions, 167
  prior and posterior, 168
Probability density function (pdf)
  bivariate normal, 219
  chi-squared, 144
  conditional, 52
  convolution, 204
  'cusp', 182
  definition, 47
  'distorted U', 73, 108
  entropy, 188

inverse chi-squared, 178
joint, 52
logarithmic, 58–61
marginal, 52
model prior pdf, 201
moments, 68
multivariate normal, 144
normal (Gaussian), 54–56, 103–104, 192
posterior, 169
prior, 169
quantities prior, 201
rectangular, 58, 105
standard normal, 55
Student's-t, 105–107, 109, 113, 177
trapezoidal, 56–58, 104–105, 205
triangular, 58, 105
truncated normal, 191

Quality, 25, 44
Quantities
adimensional, 28
as random variables, 47
correlated, 53, 91
definition, 28
directly measured, 81–87, 91
general, 28
imported, 40, 77, 91
independent, 53, 91, 116
influence, 27, 84, 86
input, 40
linearized, 63, 114
output, 40
particular, 28
primary, 40, 77, 91, 127
prior pdf, 201
reproduced, 30, 93
tabulated, 77–79
test, 94, 131, 155

Random
effects, 132, 176
error, 36, 56

experiment, 47
sample, 82, 111
scatter, 86
sequence, 111
variable, 47
Reactance, 127
Recommendation INC-1, 45–46
Reference material, 30, 93
Repeatability, repeatable measurements, 33, 82, 111, 158
Reproducibility, reproducible measurements, 32, 43, 149–150, 184–187
Resistance, 127, 147
Resistor, 31, 93, 95, 104, 129
Resolution, 32, 86, 158, 182
Risks, producer's and consumer's, 109–111, 192–201

Sample
mean, 83, 111
space, 47
variance, 83, 85, 111
Sampling theory, 82, 117
Scale marks, 87, 93
Seebeck voltage, 150
Significant figures, 79, 81
Software packages, 122
Specifications, 26, 109, 193
Standard (of measurement), 30, 92–101, 131, 155
Standard deviation, 51
Standard normal cumulative function, 55
Statistic, 83
Systematic
effects, 36, 88, 96, 155
error, 88

Tare weight, taring, 94, 101, 155
Taylor series expansion
first order, 62–63
second order, 67–68

Temperature, 33, 79–81, 88–92, 142, 150–154
Test quantity, 94, 131, 155
Test uncertainty ratio (TUR), 201
Thermocouple, 150
Thermometer, 142, 151
Tolerance, 26, 109
t-quantiles, 106
Traceability, 29, 32, 34
Type A (TA) analysis, 81–85, 111–114
Type B analysis, 83, 115

Uncertainty
  calibration, 154, 165
  combined, 66
  components, 64
  coverage, 107–108, 192
  definition, 46
  enlarged, 51, 145
  expanded, 53, 103, 110, 193
  general concepts, 32–35, 41–44
  matrix, 123, 127
  mutual, 53
  random and systematic, 38, 45
  relative standard, 51
  relative uncertainty of, 116
  reliability, 115–116
  repeatability component, 85
  standard, 51
  standard *versus* expanded, 117–118
  true, 43, 117
  uncertainty of, 115, 117, 178
  user's, 154, 165

Unit (of measurement)
  decibel, 72
  definition, 28
  symbol, 28

Value
  adjusted, 134
  adopted, 149
  best estimated, 51
  conventional true, 31, 36, 93, 155, 161
  expected, 49
  nominal, 26, 31, 96, 109–110, 161, 193
  numerical, 28
  of a quantity, 28
  of the unit, 28
  reference, 149
  true, 36, 43, 88, 109
Variance, 50
Vector
  definition, 120
  of functions, 138
  of input and output quantities, 123
Verification, 98–99, 160, 198
Volume, 88–92

Weighing instrument, 163
Welch–Satterthwaite (WS) formula, 115
Wrist watches, 101

Zeroing, 89, 100